AF806299

Tracking Control of Networked Systems via Sliding-Mode

Meng Li · Yong Chen · Ikram Ali

Tracking Control of Networked Systems via Sliding-Mode

Springer

Meng Li
School of Automation Engineering
University of Electronic Science
and Technology of China
Chengdu, Sichuan, China

Yong Chen
School of Automation Engineering
University of Electronic Science
and Technology of China
Chengdu, Sichuan, China

Ikram Ali
School of Automation Engineering
University of Electronic Science
and Technology of China
Chengdu, Sichuan, China

ISBN 978-981-16-6513-4 ISBN 978-981-16-6514-1 (eBook)
https://doi.org/10.1007/978-981-16-6514-1

© The Editor(s) (if applicable) and The Author(s), under exclusive license to Springer Nature Singapore Pte Ltd. 2022
This work is subject to copyright. All rights are solely and exclusively licensed by the Publisher, whether the whole or part of the material is concerned, specifically the rights of translation, reprinting, reuse of illustrations, recitation, broadcasting, reproduction on microfilms or in any other physical way, and transmission or information storage and retrieval, electronic adaptation, computer software, or by similar or dissimilar methodology now known or hereafter developed.
The use of general descriptive names, registered names, trademarks, service marks, etc. in this publication does not imply, even in the absence of a specific statement, that such names are exempt from the relevant protective laws and regulations and therefore free for general use.
The publisher, the authors and the editors are safe to assume that the advice and information in this book are believed to be true and accurate at the date of publication. Neither the publisher nor the authors or the editors give a warranty, expressed or implied, with respect to the material contained herein or for any errors or omissions that may have been made. The publisher remains neutral with regard to jurisdictional claims in published maps and institutional affiliations.

This Springer imprint is published by the registered company Springer Nature Singapore Pte Ltd.
The registered company address is: 152 Beach Road, #21-01/04 Gateway East, Singapore 189721, Singapore

Acknowledgements

This work was supported by the National Natural Science Foundation of China under Grant (61903064 and 61973331), the Key Research and Development Program of Sichuan Province under grant (2021YFG0205 and 2021YFG0079), the China Postdoctoral Science Foundation Funded Project under grant (2019M663479), and the Fundamental Research Funds for the Central Universities under grant (ZYGX2019J058).

Contents

Chapter 1
Related Background and Pre-knowledge

The network control system (NCS) has become a hot research topic in academia and industry. However, due to an imperfect communication structure, network node topology changes, external disturbances, and hacker attacks, the NCS faces numerous problems and challenges. In this article, we review some recent challenging research on NCS. Several related NCS research topics are investigated, including network-induced latency, packet loss, packet clutter, external interference, parameter uncertainty, sensor or actuator failure, scheduling protocols, network attacks, and other topics. The control laws to deal with these problems and potential future research directions are discussed.

Since the end of the twentieth century, with the advent of the age of information technology, computer and communication technology have been developed rapidly. In addition, communication networks have been introduced into control systems; thus, NCS emerged [106]. Research on NCS can be divided into two categories [27]. The first is the control of the network related to the study of communications and methods to ensure that the network is suitable for real-time NCS. The second is the control of the network related to control policies and control system design of the network to mitigate the adverse effects of network parameters, such as latency and packet loss. This paper focuses on the latter.

An NCS is a feedback control system whose system components (such as sensors, controllers, and actuators) are distributed in different locations and connected via wired/wireless communication networks, as shown in Fig. 1.1, depicting the typical structure of an NCS. The solid and dashed lines indicate direct transmissions and transmissions over the network, respectively. Other processes, such as data caching and parameter estimation, may be involved in the feedback process. Compared with traditional point-to-point control systems, an NCS has the advantages of resource sharing, low cost, easy expansion, simple installation and maintenance, and high reliability ([92]; Zhang 2018; [159]). Therefore, in recent years, NCSs have been widely deployed in various fields, such as cloud control systems [124, 126], smart grids [45, 46], big data [35], remote operating systems [141], and fault detection systems [55, 117].

© The Author(s), under exclusive license to Springer Nature Singapore Pte Ltd. 2022

M. Li et al., *Tracking Control of Networked Systems via Sliding-Mode*,
https://doi.org/10.1007/978-981-16-6514-1_1

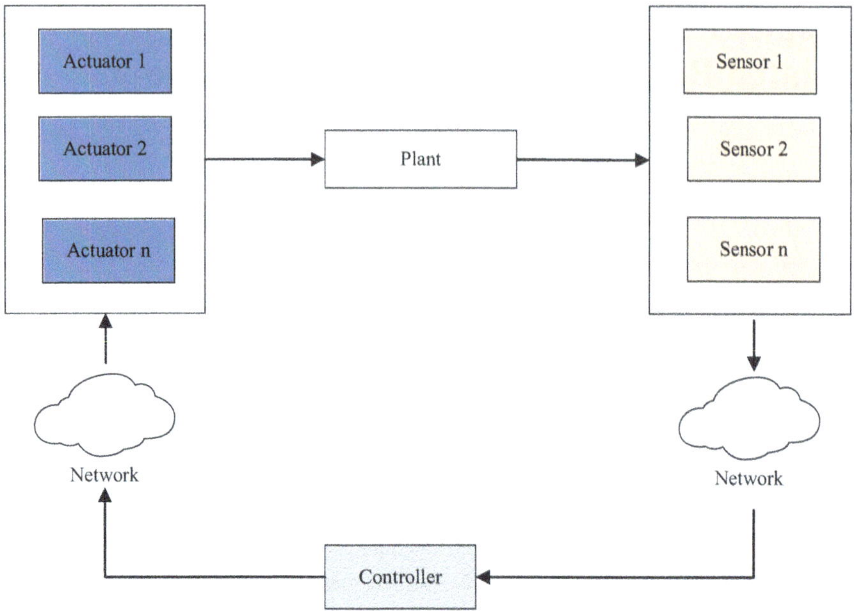

Fig. 1.1. Typical structure of NCSs

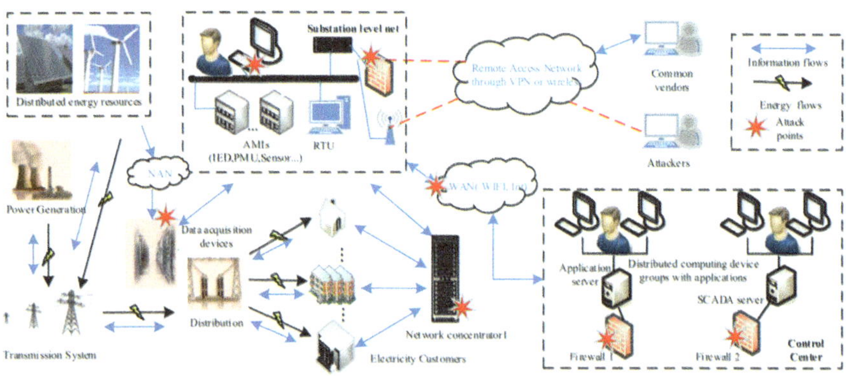

Fig. 1.2. Industrial applications of NCSs

Figure 1.2 shows some industrial applications of an NCS and some cross-disciplinary impacts that NCS promotes. First, NCSs have significantly promoted the development of modern industry. Second, due to imperfect communication, node movement, external disturbances, and hacker attacks [5, 12, 25, 39, 42, 72, 87, 88, 91, 148, 171], traditional control methods of NCSs have resulted in new challenges. These challenges may invalidate existing control algorithms or degrade system performance, causing difficulties for researchers and leading to economic

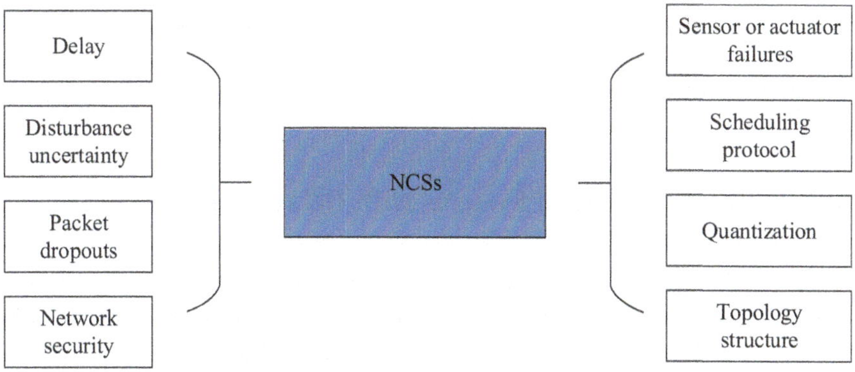

Fig. 1.3 Research subjects in NCSs

losses. To date, most existing works on NCS have focused on how to compensate and overcome the adverse effects caused by these problems.

Several investigative articles have been published regarding NCS. In Xia et al. [127], the latest advances in NCS were investigated, and the results of quantification, estimation, fault detection, and network predictive control (NPC) were discussed. A cloud control system was proposed. Gupta and Chow [27] discussed the history and evolution of NCS and analyzed network technologies, network resource allocation, network latency, scheduling, network security, and fault tolerance. Zhang et al. [158] investigated the main methods to deal with latency, chaos, packet loss, multi-node access network competition, time-varying transmission interval, and data quantization. Zhang et al. [153] analyzed the sampling data control and event triggering control in NCSs and summarized existing NCS research methods.

This paper differs from existing reviews in that it focuses on the latest progress in the analysis and design of NCSs. More research topics than in previous reviews are presented, including network-induced latency, packet loss, packet clutter, interference, uncertainty, sensor and actuator failures, scheduling protocols, network attacks, and more. A mathematical model of an NCS under various constraints is presented. The control strategies to solve these problems are discussed, and the characteristics of each method are analyzed. Potential future research directions are explored. Figure 1.3 shows the main NCS research topics. Most studies only considered some of these topics and focused on ideal conditions. It is crucial to investigate all these NCS topics and incorporate the latest knowledge and information to keep up with current research.

The rest of this chapter is organized as follows. In the first part, we discuss network-induced latency and packet loss. More specifically, the reasons for delay and packet loss are analyzed, and various control strategies are introduced. The second part discusses the effects of uncertainty and disturbances on NCSs. In the third part, the faults of the sensor and actuator are discussed. The fourth part analyzes the network scheduling protocol, and the fifth part focuses on network attacks and security problems. The sixth part provides the summary and future research prospects.

Notations Throughout the chapter, $\|g\|$ denotes the Euclidean norm. $\Pr\{\bullet\}$ is the probability of the event. Once defined, variables retain their meaning. In addition, matrices have appropriate dimensions unless otherwise stated.

1.1 Networked Control Systems with Delays and Packet Dropouts

Over the past few decades, network-induced latency and packet loss have been of significant concern in industrial control. One of the main reasons is that they usually lead to a decline in the dynamic performance of the system or even to system instability [152].

1.1.1 Problem Formulation and Modeling

Since there are many information sources in an NCS, when nodes exchange data over the network, unavoidable data conflicts, multipath transmission, link interruptions, network congestion, and other phenomena will lead to network-induced delays [107, 129]. As shown in Fig. 1.4, $\overline{\tau}^{sc}$ and $\overline{\tau}^{ca}$ represent the delay in the feedback channel (from the sensor to the controller) and the feedforward channel (controller to the actuator), respectively. In general, delays can be divided into three categories: constant, time-varying, and random. In particular, the latter two conditions are widespread in real systems. Therefore, scholars have proposed many control methods to deal with them.

Constant delay The following discrete-time NCS has been widely researched [152]:

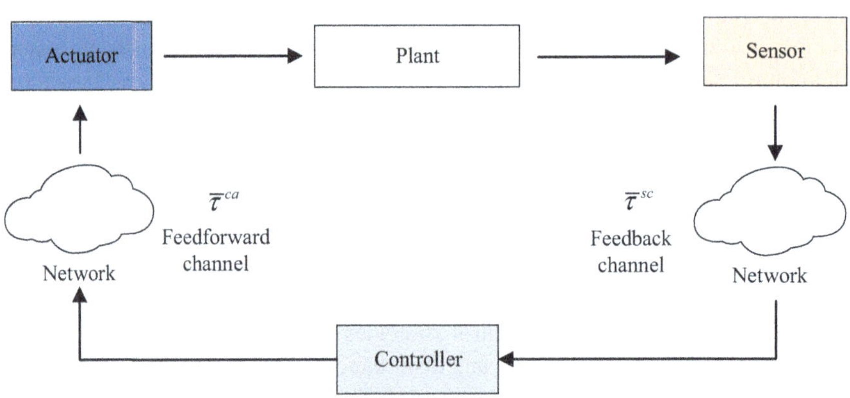

Fig. 1.4 Delay and packet dropouts in NCSs

$$\begin{aligned} \overline{x}_{k+1} &= \overline{A}\overline{x}_k + \overline{B}u_k \\ \overline{y}_k &= \overline{C}x_k \end{aligned} \tag{1.1}$$

where $\overline{x}_k \in R^n$, $u_k \in R^m$ and $\overline{y}_k \in R^l$ are the state, input, and output vectors of the system, respectively, and $\overline{A} \in R^{n\times n}$, $\overline{B} \in R^{n\times m}$ and $\overline{C} \in R^{l\times n}$ are the system matrices with appropriate dimensions.

Zhang et al. [152] proposed NPC to compensate for the influence of constant network delay, considering the delay caused by the entire network in the communication network. The main function of NPC is to predict the future behavior of the system and take corresponding control measures based on the real-time information of the delay. The Takagi–Sugeno fuzzy system was used in the analysis. The ith principle of the fuzzy system has the following format:

Rule i: If $\overline{\theta}_1(t)$ is $\overline{F}_{i1}$, and $\overline{\theta}_n(t)$ equals $\overline{F}_{in}$, then

$$\begin{aligned} \dot{\overline{x}}(t) &= \overline{A}_i\overline{x}(t) + \overline{A}_{di}\overline{x}(t-d) + \overline{B}_iu(t) \\ \overline{y}(t) &= \overline{C}_i\overline{x}(t) \end{aligned} \tag{1.2}$$

where $\overline{\theta}_i(t)$ $(i = 1, \ldots, n)$ are premise variables, $\overline{F}_{ij}$ (j = 1, . . . ,n) are the fuzzy set, and d is the delay. Zhang et al. [152] designed a cost-assured network control method to overcome the influence of constant time delay. The cost control depends on the cost function to achieve optimal control. The optimal controller is obtained by minimizing the random cost function.

In Matiakis et al. [70], the following linear time-invariant (LTI) system is analyzed:

$$\overline{G}_p = \overline{Y}_p(s)/\overline{U}_p(s) \quad \text{and} \quad \overline{G}_c = \overline{Y}_c(s)/\overline{U}_c(s) \tag{1.3}$$

where $\overline{Y}_p(s)$, $\overline{U}_p(s)$, $\overline{Y}_c(s)$, and $\overline{U}_c(s)$ are the Laplace transforms of the input and output of the plant and the controller. The constant delay is considered in the feedback channel. Under the influence of a constant delay, scattering transformation is used to ensure the stability of the system.

Cho and Fadali [10] analyzed the following nonlinear systems with constant delay:

$$\dot{\overline{x}}(t) = f(\overline{x}, t) + g(\overline{x}, t)u(t - \overline{\tau}_u) \tag{1.4}$$

where $f(\bullet)$ and $g(\bullet)$ are nonlinear functions, and $\overline{\tau}_u$ represents delay. Reset control and a neural network are used to compensate for errors due to constant delays. Tan and Liu [102] studied the consensus problem of networked multi-agent systems with constant delay in communication channels and proposed a novel distributed protocol to compensate for the constant delay.

Time-varying delay Kima and Park [36] used the following state-space equation to investigate time-varying delays:

$$\dot{\overline{x}}(t) = \overline{A}\overline{x}(t) + \overline{A}_d\overline{x}(t - \overline{d}_1(t)) + \overline{B}_1u(t) + \overline{B}_du(t - \overline{d}_2(t)) \tag{1.5}$$

where $\overline{d}_1(t)$ and $\overline{d}_2(t)$ are two time-varying delays of the state and the input. The delay was considered the interference of the system, and a robust state feedback controller was designed to mitigate the effect of the delay. The robust control approach considered the delay as a multiplicative perturbation of the system. A robust control was used to minimize the influence of the perturbation and maintain the system performance. Here, the multiplicative perturbation refers to the delay acting on the transfer function in the form of a product. The advantage of this method is that the controller does not need a priori information about the network delay probability distribution.

Zhu et al. [169] analyzed the following discrete-time switching NCS with a time-varying delay:

$$\overline{x}(k+1) = \overline{A}_{\overline{\sigma}(k)}\overline{x}(k) + \overline{B}_{\overline{\sigma}(k)}f(\overline{x}(k)) + \overline{C}_{\overline{\sigma}(k)}g(\overline{x}(k - \overline{\varphi}_{\overline{\sigma}(k)}(k))) \tag{1.6}$$

where $f(\bullet)$ and $g(\bullet)$ are two nonlinear functions, $\overline{\varphi}_{\overline{\sigma}(k)}(k)$ represents the delay change over time, and $\overline{\sigma}(k)$ are piecewise constant functions of time k, which are evaluated in a finite set S. A control algorithm based on the Lyapunov–Krasovskii function was designed, and it was found that the delay-dependent condition was valid. The linear matrix inequality (LMI) technique and Schur's complement lemma were used to solve the controller gain.

Khanesar et al. [34] analyzed the following nonlinear system:

$$\begin{cases} \dot{\overline{x}}_1 = \overline{x}_2 \\ \vdots \\ \dot{\overline{x}}_{n-1} = \overline{x}_n \\ \dot{\overline{x}}_n = f(\overline{X}) + g(\overline{X})u(t) + \overline{d}(t) \end{cases} \tag{1.7}$$

where $x_i (i = 1, 2, \ldots, n)$ is the state, $\overline{X}$ is the state vector of the system, $f(\cdot)$ and $g(\cdot)$ are two nonlinear functions, and $\overline{d}(t)$ represents the interference signal. The time-varying delay in communication networks was considered, and a novel fuzzy sliding mode controller was designed to compensate for the delay.

Zhang et al. [149] described a controlled system with time-varying delays

$$\begin{cases} \overline{x}(k+1) = \overline{A}_{\overline{\sigma}(k)}\overline{x}(k) + \overline{B}_{\overline{\sigma}(k)}f(\overline{x}(k)) + \overline{C}_{\overline{\sigma}(k)}g(\overline{x}(k - d(k))) \\ \overline{y}(k) = \overline{E}_{\overline{\sigma}(k)}\overline{x}(k) \\ \overline{z}(k) = \overline{G}_{\overline{\sigma}(k)}\overline{x}(k) \end{cases} \tag{1.8}$$

where $f(\cdot)$ and $g(\cdot)$ denote neuron activation functions; $\overline{\sigma}(k)$ is the switching signal; $d(k)$ presents the time-delay; $\overline{x}(k)$, $\overline{y}(k)$ and $\overline{z}(k)$ denote the state, measure output, and observed output; $\overline{A}_{\overline{\sigma}(k)}$, $\overline{B}_{\overline{\sigma}(k)}$, $\overline{C}_{\overline{\sigma}(k)}$, $\overline{E}_{\overline{\sigma}(k)}$ and $\overline{G}_{\overline{\sigma}(k)}$ are coefficient matrices. A mode-dependent estimator was designed to deal with the delays.

Wang et al. [108] proposed NPC based on H∞ control for a class of NCSs with time-varying delays in the forward and backward channels to compensate for the delay in the two channels. Cloosterman et al. [11] studied uncertain, time-varying network delays in an NCS and developed a stability analysis and controller synthesis for plants. Zhang et al. [158] investigated the system stability of NCSs with time-varying network communication delays.

Random delays Zhang et al. [154] analyzed a class of linear discrete-time systems with random delays for the states, control, and output vectors as follows:

$$\overline{x}(k+1) = \sum_{i=0}^{\overline{\tau}_a} \overline{A}_i \overline{x}(k-i) + \sum_{j=0}^{\overline{\tau}_b} \overline{B}_j u(k-j) \tag{1.9a}$$

$$\overline{y}(k) = \sum_{h=0}^{\overline{\tau}_c} \overline{C}_h \overline{x}(k-h) \tag{1.9b}$$

where $\overline{\tau}_a$, $\overline{\tau}_b$ and $\overline{\tau}_c$ are the upper bounds of the delay. A model-free optimal control scheme was proposed to address the delays.

Hu et al. [31] used an autoregressive moving-average model described as:

$$\overline{A}(\overline{z}^{-1})\overline{y}(t) = \overline{B}(\overline{z}^{-1})u(t-1) \tag{1.10}$$

where $\overline{A}(\overline{z}^{-1})$ and $\overline{B}(\overline{z}^{-1})$ are two polynomials. Random network delays were assumed to reduce the control performance and cause system instability. A new event-driven NPC method was proposed to compensate for the random delays.

Shi et al. [96] performed stochastic delay modeling using Markov chains to achieve robust tracking control for NCSs. The transition probability information between different delay elements was required. Probst et al. [83] used the Pade approximation method to model random time delays in the frequency domain and performed estimations using a Kalman filter. Yang et al. [134] evaluated the stochastic communication delay in NCSs and analyzed the predictive output feedback control problem. A networked predictive control scheme was adopted to compensate for delay caused by the random network. Liu et al. [56] used predictive control for an NCS with random network delay in the feedback channel to minimize the delays and provided a stability standard of the closed-loop system. NPC can actively compensate for delays. Zhang et al. [158] investigated stepping tracking control for discrete-time nonlinear network systems with random delay. Zhang et al. [156] proposed a class of NCSs in a discrete-time domain with random delays. The sensor-to-controller delay and controller-to-actuator delay were modeled as two Markov chains, and the stability of the closed-loop system was analyzed.

The control methods described by Kima and Park [36], Zhu et al. [169], Matiakis et al. [70], Shi et al. [96], Probst et al. [83], and Zhang et al. [154] are basically by passively accepting the existence of time-delay caused by the network. It is worth noting that the robust control method described in the literature does not require redesigning the existing predictive controller or clock synchronization. When the controlled system has uncertain parameters and external disturbances, the robust H-infinite control and Kalman filter provide effective control. When the model information is known, and the delay limit is given, the NPC method can achieve good control performance, as shown in the references [31, 56, 108, 134, 152, 162].

1.1.2 Packet Dropouts

Node failures and information conflict can occur during network transmission because of message conflict and node contention for bandwidth. When a packet transmission fails or is not received within the specified time, the packet is discarded [161]. In Fig. 1.4, switch describes packet loss, where "off" means a lost transmission, and "on" means a successful transmission. In general, there are three strategies for dealing with data loss.

(1) The controller and actuator use the previous control input, i.e., the hold-input strategy. As shown in Fig. 1.5, the input is $u(k)$ at the time k; a packet is lost at time $k+1$. Hence, $u(k)$ is held until $k+2$. No packet dropouts are occurring at times $k+2$ and $k+3$, and the inputs are $u(k+2)$ and $u(k+3)$, respectively. However, packet dropouts occur at times $k+4$ and $k+5$. Therefore, $u(k+3)$ is held until time $k+6$.

(2) The controller and actuator assume a zero input, i.e., the zero–input strategy.

(3) The controller and actuator use predictive control as input, i.e., a predictive control strategy. Strategies 1 and 2 are simple, but they may not be straightforward to implement and can be difficult to use to achieve the required control performance. Strategy 3 can achieve the desired control performance, but a predictive control method must be designed. Lin et al. [52] used the following control strategy that considered packet dropouts:

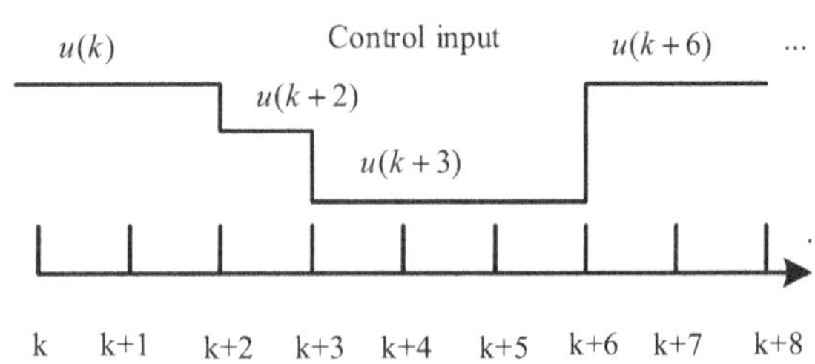

Fig. 1.5. The process of packet transmission under the hold-input strategy

$$\begin{cases} \overline{x}_{k+1} = \overline{A}\overline{x}_k + \overline{\gamma}_k \overline{B} u_k \\ \overline{y}_k = \overline{C}\overline{x}_k \end{cases} \tag{1.11}$$

where $\overline{\gamma}_k$ describes the packet losses in the controller-actuator channel., i.e.,

$$\overline{\gamma}_k = \begin{cases} 1, u_k \text{ has been successfully delivered to actuator,} \\ 0, \text{ otherwise.} \end{cases}$$

The optimal estimation problem of random discarded control packets in lossy networked control systems is studied.

Several studies focused on estimating randomly discarded control packets in lossy NCSs. Shen et al. [94] solved the iterative learning control problem of a linear NCS with random data loss using modeling with Bernoulli random variables. A random variable $\overline{\gamma}_k$ was introduced to indicate whether the measured packet $\overline{y}_k(t)$ had been successfully sent

$$\overline{\gamma}_k(t) = \begin{cases} 1, \ \overline{y}_k(t) \text{ is successfully transmitted,} \\ 0, \text{ otherwise.} \end{cases}$$

with the probability distribution:

$$P(\overline{\gamma}_k(t) = 1) = \overline{\rho} \text{ and } P(\overline{\gamma}_k(t) = 0) = 1 - \overline{\rho} \text{ with } 0 < \overline{\rho} < 1 \tag{1.12}$$

Xu et al. [129] described a nonlinear discrete-time Markov jump system using the following T–S fuzzy model.

Plant Rule i: If $\overline{\mathrm{Q}}_1$ is $\overline{\delta}_{\mathrm{i1}}$, and $\overline{\mathrm{Q}}_\mathrm{p}$ equals $\overline{\delta}_{\mathrm{ip}}$, then

$$\begin{cases} \overline{x}(k+1) = \overline{A}_{i\overline{\tau}(k)}\overline{x}(k) + \overline{B}_{i\overline{\tau}(k)}u(k) \\ \overline{z}(k) = \overline{C}_{i\overline{\tau}(k)}\overline{x}(k) + \overline{D}_{i\overline{\tau}(k)}u(k) \end{cases} \tag{1.13}$$

where $\overline{\tau}(k)$ is the Markov random variable, and the transition probability matrix $\overline{\Omega} = \{\pi_{ij}\}$ is given by

$$P\{\overline{\tau}(k+1) = j | \overline{\tau}(k) = i\} = \pi_{ij} \text{ with } 0 < \pi_{ij} < 1 \tag{1.14}$$

The Bernoulli process also describes a nonlinear discrete-time Markov jump factory with packet loss. A non-fragile asynchronous controller was designed to overcome the packet loss problem.

Peng et al. (2015) discussed a nonlinearly controlled factory as follows:

$$\overline{y}(k+1) = f(\overline{y}(k), \ldots, \overline{y}(k-n_y), u(k), \ldots, u(k-n_u)) \tag{1.15}$$

where n_y and n_u are the unknown orders of the output and input, respectively. For a networked nonlinear system (1.15) with two-channel packet dropouts, a data-based predictive control was proposed to compensate for the packet dropouts.

Zhang et al. [152] analyzed NCSs with packet loss. The output feedback was used, and the stability of the controlled plant was discussed. Wang et al. [118] analyzed feedback passivity for nonlinear systems with packet loss.

It may be noted from the references that control methods designed to solve latency can also be used for packet loss problems. In some references, packet loss is referred to as an increase in latency. For example, if the sampling period is T, and the number of continuously lost packets is d, the delay caused by the packet loss is $\overline{\tau}_d = dT$.

It is worth mentioning that the above studies only focused on delays caused by the network or only analyzed the effect of packet loss in the network. However, in real network traffic, both latency and packet loss occurs in the NCS.

1.1.3 Simultaneous Delays and Packet Dropouts

Tan et al. [101] analyzed the following controlled plant that experienced both delays and packet dropouts:

$$\overline{x}_{k+1} = \overline{A}\overline{x}_k + \overline{\gamma}_k \overline{B} u_{k-d} \tag{1.16}$$

where d is the delay and $\overline{\gamma}_k$ is the packet dropout index. Tan and others investigated simultaneous network-induced delay and packet loss in discrete-time NCSs. Their results show that the maximum packet loss rate can be explicitly calculated if the network-induced latency is known in advance. If the grouping loss rate is known in advance, the maximum allowable delay limit can also be determined.

Xua et al. [131] evaluated the following controlled plant that experienced delays and packet dropouts:

$$\dot{\overline{x}}(t) = \overline{A}\overline{x}(t) + \overline{\gamma}(t)\overline{B}u(t - \overline{\tau}(t)) \tag{1.17}$$

where $\overline{\tau}(t)$ is the time-varying delay and $\overline{\gamma}(t)$ is the packet dropout index. Stochastic optimal control was analyzed, and the optimal controller was derived by minimizing the following stochastic cost function:

$$\overline{J}_k = \underset{\overline{\tau},\overline{\gamma}}{E}\left[\sum_{m=k}^{\infty} (\overline{x}_m^T \overline{S}\overline{x}_m + u_m^T \overline{R} u_m)\right] \quad k = 0, 1, \ldots \tag{1.18}$$

Rakkiyappan et al. [91] analyzed the stability of NCSs with control packet loss and additive time-varying delays of a plant with N identical coupled nodes, where each node was an n-dimensional dynamical system. The ith node was described as follows:

$$\dot{\overline{x}}_i(t) = f(\overline{x}_i(t)) + c\sum_{j=i}^{N} \overline{G}_{ij}\overline{A}\overline{x}_j(t - \overline{d}_1(t) - \overline{d}_2(t)) + u_i(t) \tag{1.19}$$

where c is a constant denoting the coupling strength, $\overline{G}_{ij}$ is the outer coupling configuration matrix representing the topological structure of the network, A denotes the gain matrix, $u_i(t)$ stands for the control input, and $\overline{d}_l(t)(l = 1, 2)$ are the two time-varying delays.

Wang et al. [118] considered the following controlled plant:

$$\begin{cases} \overline{x}(k+1) = \overline{A}(k)\overline{x}(k) + f(\overline{x}(k), \overline{\xi}(k)) \\ \overline{y}(k) = \overline{C}(k)\overline{x}(k) \end{cases} \tag{1.20}$$

where $f(\bullet)$ is a nonlinear function, $\overline{\xi}(k)$ is a zero-mean Gaussian noise sequence, and represents the stochastic nonlinearity of the state. For example, $f(\overline{x}(t), \overline{\xi}(t)) = sin(\overline{x}_{t1})\overline{\xi}_{t1} + \cos(\overline{x}_{t2})\overline{\xi}_{t2}$. Recursive estimation was conducted for a discrete time-varying plant (1.19) with multi-step transmission delays and multiple packet dropouts.

Quevedo and Jurado [90] assessed the stability of sequence-based control for the following nonlinear plant with random delays and packet dropouts in the communication channels:

$$\overline{x}(k+1) = f(\overline{x}(k), u(k)) \tag{1.21}$$

Li and Chen [43] focused on random delays and packet loss in the feedback channel and the feedforward channel of a DC permanent magnet motor in an NCS. A sliding mode predictive tracking control algorithm based on pseudo partial differential theory was designed to compensate for the random delays and packet loss during motor control. Donkers et al. [14] analyzed the stability of NCS-controlled factories with time-varying transmission intervals, time-varying transmission delays, and packet loss. The transmission interval and transmission delay were described by several continuous random variables. Franze et al. [18] analyzed the effect of time-varying delays and data loss on NCS stability. Zhang et al. (2008) proposed a new linear switching system model for network-induced delay and packet loss in an NCS. Sufficient conditions were derived to achieve exponential stability of a closed-loop NCS. Wang and Han [119] established a new model for a continuous-time NCS with network-induced delay and packet loss. The authors used a delay compensation method based on linear estimation for the controller design. Li et al. [40] established a sliding mode predictive controller to compensate for the time delay and packet loss in an NCS using a multi-packet transmission strategy. The Kalman predictor was used to estimate the current state and predict the future plant state, and the sliding mode control (SMC) was used to calculate the control sequence.

It is well known that network-induced latency and packet loss are the two main causes of low control performance and destabilization of the system. It is worth mentioning that several control methods have been proposed to deal with these problems.

1.2 Networked Control Systems with Disturbance

This section discusses plant uncertainties and disturbances in NCSs. Uncertainty is common in NCS due to temperature fluctuations, the gradual aging of system components, and the movement of operating points. Uncertain networked control systems (UNCS) have received widespread attention in recent years [161]. In practice, controlled plants are inevitably affected by noise and external disturbances.

Yue et al. [138] considered the following UNCS:

$$\begin{cases} \dot{\overline{x}}(t) = [\overline{A} + \Delta\overline{A}(t)]\overline{x}(t) + [\overline{B} + \Delta\overline{B}(t)]\mathrm{u}(t) + \overline{B}_w\overline{w}(t) \\ \overline{z}(t) = \overline{C}\overline{x}(t) + \overline{D}u(t) \end{cases} \tag{1.22}$$

where $\overline{w}(t)$ is the external perturbation, and $\Delta\overline{A}(t)$ and $\Delta\overline{B}(t)$ denote the parameter uncertainties. They satisfy the following condition:

$$[\Delta\overline{A}(t)\ \Delta\overline{B}(t)] = \overline{GF}(t)[\overline{E}_a\overline{\mathrm{E}}_\mathrm{b}] \tag{1.23}$$

where $\overline{G}$, $\overline{E}_a$ and $\overline{\mathrm{E}}_\mathrm{b}$ are constant matrices with appropriate dimensions, and $\overline{F}(t)$ is an unknown time-varying matrix satisfying:

$$\overline{F}^T(t)\overline{F}(t) \leq I \tag{1.24}$$

The delay and data loss caused by the network were considered, and a robust H-infinity controller was designed. The LMI technique and Schur complement lemma were used to solve the controller gain.

Yin et al. [135] developed the problem that NCSs with multiple communication channels (MCCs) are subject to energy limit interference and norm limit uncertainty. A robust controller was designed to ensure the robustness of the closed-loop system and to obtain the disturbance attenuation exponent. Communication channel switching was modeled by a Markov chain, and the transition probability from channel i to channel j was approximated as follows:

$$\Pr(\mathrm{ch}_{t+\Delta t} = j|\mathrm{ch}_t = i) = \begin{cases} \overline{\lambda}_{ij}\Delta + o(\Delta),\ \mathrm{j} \neq \mathrm{i} \\ 1 + \overline{\lambda}_{ij}\Delta + o(\Delta),\ \mathrm{j} = \mathrm{i} \end{cases} \tag{1.25}$$

where ch denotes channel, and $o(\Delta)$ is an infinitesimal of higher-order about Δ.

Wu et al. [122] investigated the following linear discrete-time UNCS:

$$\begin{cases} \overline{x}_{k+1} = (\overline{A} + \Delta\overline{A}_k)\overline{x}_k + \overline{B}u_k \\ \overline{y}_k = \overline{C}\overline{x}_k \end{cases} \tag{1.26}$$

The network time-varying delays and packet dropouts were considered, and the closed-loop NCS with uncertainties under the networked predictive controller was described as a coupled switched control system.

Qiu et al. [88], Du et al. [15], and Wang et al. [109] examined two uncertain discrete-time plants that had the following forms, respectively:

$$\begin{cases} \overline{x}_{k+1} = (\overline{A} + \Delta\overline{A})\overline{x}_k + \overline{B}u_k \\ \overline{y}_k = \overline{C}\overline{x}_k \end{cases} \tag{1.27}$$

$$\begin{cases} \overline{x}_p(k+1) = (\overline{A} + \Delta\overline{A})\overline{x}_p(k) + (\overline{B} + \Delta\overline{B})u(k) \\ \overline{y}(k) = \overline{C}\overline{x}_p(k) \end{cases} \tag{1.28}$$

Random network packet dropouts and time delays were considered, and the cost control of the output feedback was investigated.

Yan et al. [133] analyzed the following NCS with time-varying delays, parameter uncertainties, and disturbances:

$$\begin{cases} \overline{x}(t) = (\overline{A} + \Delta\overline{A}(t))\overline{x}(t) + (\overline{A}_d + \Delta\overline{A}_d)\overline{x}(t - \overline{\tau}_{sc}(t) - \overline{\tau}_{ca}(t)) + \overline{B}_w\overline{w}(t) \\ \overline{z}(t) = \overline{C}\overline{x}(t) + \overline{C}_d\overline{x}(t - \overline{\tau}_{sc}(t) - \overline{\tau}_{ca}(t)) + \overline{D}_w\overline{w}(t) \end{cases} \tag{1.29}$$

An improved robust H-infinity control was designed.

Wang et al. [118], examined the following multi-input multi-output (MIMO) UNCSs:

$$\begin{cases} \overline{x}(t) = \overline{A}\overline{x}(t) + \overline{B}u(t) + f(t, \overline{x}) \\ \overline{y}(k) = \overline{C}\overline{x}(t),\ \overline{x}(0) = \overline{x}_0 \end{cases} \tag{1.30}$$

where $f(t, \overline{x})$ is an unknown function that locally satisfies the Lipschitz condition, which denotes the uncertainty of the plant. Real-time $\overline{L}_1$ adaptive control was adopted for this plant (1.30).

SMC, a powerful nonlinear control technology, has been widely studied due to its insensitivity and high robustness. SMC provides good control performance, even in a system with communication limitations and external interference. Discrete-time SMC (DSMC) has been widely used in various fields as an implementation of digital systems, such as speed tracking control of a motor. Gao et al. [22] investigated a class of discrete nonlinear stochastic systems with multiplicative noise. This system can be described by the following TS fuzzy model:

Plant rules i: If $\overline{\delta}_{1k}$ is $\overline{\mathrm{M}}_{\mathrm{i1}}$, and $\overline{\delta}_{\mathrm{pk}}$ equals $\overline{\mathrm{M}}_{\mathrm{ip}}$. Then,

$$\begin{cases} \overline{x}_{k+1} = (\overline{A}_i\overline{x}_k + \overline{B}_i\overline{u}_k + \overline{C}_i\overline{v}_i) + (\overline{A}_i\overline{x}_k + \overline{C}_i\overline{v}_i)\overline{w}_i \\ \overline{y}_k = \overline{D}_i\overline{x}_k + \overline{E}_i\overline{v}_k \\ \overline{z}_k = \overline{F}_i\overline{x}_k \end{cases} \tag{1.31}$$

where $\overline{v}_k$ is the disturbance input, and $\overline{w}_i$ is Gaussian white noise. The observer-based H-infinity control problem was investigated.

Li et al. [41] studied nonlinear NCSs with packet dropouts and parameter uncertainties as follows:

Plant Rule i: If $f_1(\overline{\sigma}(k))$ is $\overline{M}_{i1}$, and $f_{\overline{\theta}}(\overline{\sigma}(k))$ equals $\overline{M}_{i\overline{\theta}}$. Then,

$$\begin{cases} \overline{x}(k+1) = \overline{A}_i\overline{x}(k) + \overline{B}_iu(k) + \overline{E}_i\overline{w}(k) \\ \overline{z}(k) = \overline{C}_i\overline{x}(k) + \overline{D}_iu(k) + \overline{F}_i\overline{w}(k) \end{cases} \tag{1.32}$$

A type-2 fuzzy model was used. The parameter uncertainty was accurately captured by the bottom and upper membership functions and the relative weight function.

Tang et al. [103] considered continuous-time nonlinear systems with uncertainties and disturbances which can be described by TS fuzzy models. A robust H-infinity design based on the fuzzy model was used. The rules of this model were expressed in the following if–then form:

Plant Rule i: If $\overline{\phi}_1$(t) is $\overline{\mathrm{W}}_1^i$, and $\overline{\phi}_p$(t) equals $\overline{\mathrm{W}}_p^i$. Then,

$$\begin{cases} \dot{\overline{x}}(t) = (\overline{A}_i + \Delta\overline{A}_i(t))\overline{x}(t) + (\overline{B}_i + \Delta\overline{B}_i(t))u(t) + \overline{B}_{wi}\overline{w}(t) \\ \overline{z}(t) = \overline{C}_i\overline{x}(t) + \overline{D}_iu(t) \end{cases} \tag{1.33}$$

where $\Delta\overline{A}_i(t)$ and $\Delta\overline{B}_i(t)$ denote the uncertainty matrices, and $\overline{w}(t)$ is the exogenous disturbance.

Geromel and Gabriel [24], Li and Sun [49], Morais and Peres [73], and Wang et al. [110] investigated perturbations in NCSs and robust H_2 and H-infinity control problems. Lu et al. [69] established a new state-space model for NCSs with random packet loss and uncertainty and designed an improved predictive tracking control model. Channel noise was analyzed in NCSs [142, 143], and optimal tracking was evaluated. Lu et al. [69] studied the cost control of a nonlinear uncertain NCS based on a fuzzy model.

Yue et al. [138], Yin et al. [135], Wu et al. [122], Tang et al. [103], and Yan et al. [133] showed that the uncertainty of plants was primarily attributed to the uncertainty of the state and control input. These are usually denoted by $\Delta\overline{A}$ and $\Delta\overline{B}$, respectively. Generally, the parameter uncertainties are assumed to satisfy the condition: $[\Delta\overline{A}(t)\ \Delta\overline{B}(t)] = \overline{GF}(t)[\overline{E}_a\mathrm{E}_\mathrm{b}]$ and $\overline{F}^T(t)\overline{F}(t) \leq I$. Robust control is widely used to solve uncertainties and disturbances in factories. However, it is difficult to satisfy this assumption in a real factory. There are other challenges associated with using robust control methods, such as complex matrix calculations and linearization of nonlinear inequalities. SMC is well suited to deal with interference, but the flutter problem must be overcome.

1.3 Networked Control Systems with Faults

In real-world controlled plants, system components such as sensors and actuators are often plagued by failures due to various unexpected phenomena (e.g., sensor and actuator failures), which can lead to severe system degradation and instability [95, 114]. Therefore, it is of theoretical significance and practical necessity to study sensor and actuator faults in NCSs [117, 136]. Reliable filter design is vital for detecting sensor and actuator failures in NCSs [64]. In addition, it is worth noting that most reliable filtering solutions to date have been based on fixed sensor failure models, which may introduce some limitations. However, in real systems, failures may occur randomly [16, 33, 98].

Li et al. [39] designed an H∞ filter for a nonlinear NCS with external interference and random sensor failure, which is characterized by a Bernoulli distribution. More precisely, the output $\overline{y}(k)$ sent by the sensors experienced random failures, which can be described as follows:

$$\overline{y}_{\overline{F}}(t_k) = \overline{\sigma}(t_k)\overline{y}(t_k) + [1 - \overline{\sigma}(t_k)]\overline{F}\overline{y}(t_k) \tag{1.34}$$

where $\overline{F}$ is the sensor failure function matrix; $\overline{\sigma}(t_k)$ is a Bernoulli distributed sequence defined by

$$\overline{\sigma}(t) = \begin{cases} 0, \text{ sensor failures happen} \\ 1, \text{ sensor failures do not happen} \end{cases}$$

Liu et al. [62] developed an SMC to address random sensor delays in NCSs. As described above, the sensor-measured output may experience a delay. Therefore, the measured output can be characterized by:

$$\overline{y}(t) = \overline{\alpha}(t)\overline{C}(\overline{r}(t))\overline{x}(t) + (1 - \overline{\alpha}(t))\overline{D}(\overline{r}(t))\overline{x}(t - \overline{d}(t)) \tag{1.35}$$

where $\overline{f}_{sk}$ and $\overline{f}_k$ represent the sensor and actuator faults, respectively. A new robust control method was proposed. More specifically, a disturbance observer was designed using DSMC to estimate the enhanced state and simultaneously compensate for the disturbance. Then, a robust controller based on a disturbance observer was proposed, and a new discrete sliding mode function and a second-order discrete adaptive sliding mode function were designed for state estimation and fault estimation.

1.4 Networked Control Systems with Scheduling Protocols

Currently, most of the NCS literature assumes that only one sensor (actuator) is available or that all sensors (or actuators) can simultaneously gain access to the communications network to send (or receive) signals. However, in a real factory, this assumption is invalid because the real-world network may be bandwidth-constrained. A transport protocol has been used to overcome this problem. In an NCS with multiple sensors (or actuators), the transport protocol or schedule is a basic communication flaw [30], which may cause data conflicts during multiple simultaneous access attempts.

There are three popular protocols for NCS: round-robin (RR) protocol [105], try-once discard (TOD) protocol [59], and stochastic communication protocol (SCP) [164]. Ugrinovskii and Fridman [105] analyzed distributed H-infinity estimation based on the time-delay system method for sensor networks using the RR protocol schedule. Zhang et al. [164] analyzed the optimal linear estimation of network systems constrained by the SCP protocol. Donkers et al. [14] used all three protocols in an NCS to analyze the stability based on the switching system approach. Freirich and Fridman [19] used the TOD and RR protocols to coordinate the transmission of sensor data to the controller. The combined protocol was expressed as:

$$\overline{J} = \overline{J}_{TOD} \cup \overline{J}_{RR} \tag{1.36}$$

where the $\overline{J}_{TOD} = \{j \in \overline{J} | j\text{th subsystem is under TOD}\}$, $\overline{J}_{RR} = \{j \in \overline{J} | j\text{th subsystem is under RR}\}$. The stability of the close-loop system under the two protocols was studied.

The SCP is a widely used model to describe a specific class of carrier-sense multiple access with collision avoidance (CSMA/CA) protocol. The CSMA/CA protocol has been implemented in various communication systems, such as the IEEE 802.11-based wireless local area network and the IEEE 802.15.4-based wireless sensor network [13, 171]. Dai et al. [13] used a random protocol and an I.D. for an NCS with packet loss or collision sequences with Bernoulli random variables (independent and uniformly distributed). Zou et al. [171] proposed an observer-based H∞ control for an NCS with a random communication protocol. SCP scheduling was considered a random process subject to a Markov distribution. If $\overline{\lambda}(k) \in \overline{S} = \{1, \ldots, N\}$ was the selected sensor node at time k, the transition probability from state $\overline{\lambda}(k) = i$ to state $\overline{\lambda}(k+1) = j$ was characterized as:

$$\Pr\{\overline{\lambda}(k+1) = j | \overline{\lambda}(k+1) = i\} = \pi_{ij} \tag{1.37}$$

where π_{ij} is the transition probability, and $0 \leq \pi_{ij} < 1$. Liu et al. [61] analyzed two classes of stochastic protocols for NCSs with variable delays. One class was characterized by an independent and identically distributed stochastic process, in which the activation probability of each sensor node was a given constant. For instance, let

$\overline{\varsigma}_k \in \overline{\Gamma} = \{0, 1, \ldots, N\}$ represent the active sensor node, which is an i.i.d with the probabilities given by:

$$\Pr\{\overline{\varsigma}_k = i\} = \overline{\alpha}_i, i \in \overline{\Gamma} \tag{1.38}$$

where $\overline{\alpha}_i$ is a non-negative scalar and $\sum_{i=0}^{N} \overline{\alpha}_i = 1$. The second class was the scheduling protocol defined by a discrete-time Markov chain with a known transition probability matrix. Yin et al. [135] investigated a class of NCSs,the active data communication channel in MCCs was switched, and the switching was governed by a Markov chain. Let $\overline{c}_t$ denote the channel-switching signal using values in a finite set $\overline{C} = \{1, 2, \ldots, N\}$, indicating the number of the current channel. The probability of transition from channel $\overline{c}_t = i$ to channel $\overline{c}_{t+\Delta t} = j$ can be approximated as:

$$\Pr\{\overline{c}_{t+\Delta t} = j | \overline{c}_t = i\} = \begin{cases} \overline{\lambda}_{ij}\Delta + o(\Delta), \ \mathrm{j} \neq \mathrm{i} \\ 1 + \overline{\lambda}_{ij}\Delta + o(\Delta), \ \mathrm{j} = \mathrm{i} \end{cases} \tag{1.39}$$

where $o(\Delta)$ satisfies $\lim_{\Delta \to 0} o(\Delta)/\Delta = 0$.

As far as the authors know, the performance of NCSs depends largely on the design of its control algorithm. Therefore, control methods and scheduling have attracted significant attention [6, 20]. Zhang and Varsakelis [157] proposed a joint design scheme to stabilize LTI systems in which a finite number of sensors/actuators exchanged information with remote controllers over a shared common communication network. Zhao et al. [167] proposed a collaborative design method of predictive control and scheduling to deal with communication constraints in a group of NCs with network-induced delay. Dai et al. [13] focused on feedback control and scheduling co-design and analyzed the uncertainty of the delay caused by the network in the control loop in the sampling data control plant. Sun et al. [99] ensured the input state stability of controlled plants using the small gain theory for NCS with Unified Global Exponentially Stable (UGES) protocol. Lješnjanin et al. [65] examined NCSs affected by packet loss and scheduling using a protocol and controller co-design approach.

1.5 Networked Control Systems with Network Security

As another research branch of NCSs, network security has attracted increasing attention from researchers. In the past few years, several cyber-attacks have attracted worldwide attention. Here, we focus on some of them. Cyber-attacks pose a serious threat to smart vehicles. Hacking attacks against smart cars are often reported. For example, in 2015, two hackers in the United States used the Internet to break into the electronic systems of a Jeep Cherokee, causing the car to crash into a roadside ditch.

In addition, in 2017, the ransomware virus caused extensive disturbances worldwide, especially on university campuses. We are often helpless in the face of cyber-attacks.

From the perspective of the control system, research on network security is still in its infancy. To date, most of the literature on cyber security issues has focused on security attacks [27, 29], as shown in Fig. 1.6. Attackers can launch various types of attacks against different components of a network attack.

Intrusion detection is a key step to defend against network attacks. Kwon and Hwang [37] proposed an algorithm to evaluate the security of network physical systems in the presence of network attacks. The algorithm has been intelligently designed to avoid detection. Pasqualetti et al. [80] designed a centralized and distributed attack detection and recognition method for cyber-physical systems (CPSs) affected by malicious attacks. Chen et al. [7] proposed a novel attack detection method for a power system by analyzing the differences in the dynamic characteristics of the variables.

Network attacks include denial of service (DoS) attacks [1] and data injection attacks [71]. Methods to prevent DoS attacks against NCS include fault tolerance and attack compensation control methods. In the former, a control policy is utilized that tolerates network-induced undesirable phenomena caused by a DoS attack. In the latter, the control methods compensate for the loss of control performance caused by a DoS attack by contradicting the phenomenon caused by an unwanted attack. Li et al. [50] developed a data sending strategy to prevent packet loss caused by a DoS attack. Amin et al. [2] used the concept of safe investment for elastic control. Unfortunately, to the best of the authors' knowledge, all these elastic control schemes have been developed in continuous or discrete domains,thus, potential numerical problems may be ignored for sampling systems with high sampling rates. Therefore, Yuan et al. (2013) proposed a new elastic control method for DoS attacks on NCSs using incremental operators. An and Yang [4] developed an improved adaptive elastic control scheme to mitigate data injection attacks in network physical systems. The effects of the unknown sensor and actuator attacks were successfully mitigated by using the two-step backward method, the adaptive boundary estimation mechanism, and the fast growth rate of the Nussbaum function. Pang et al.

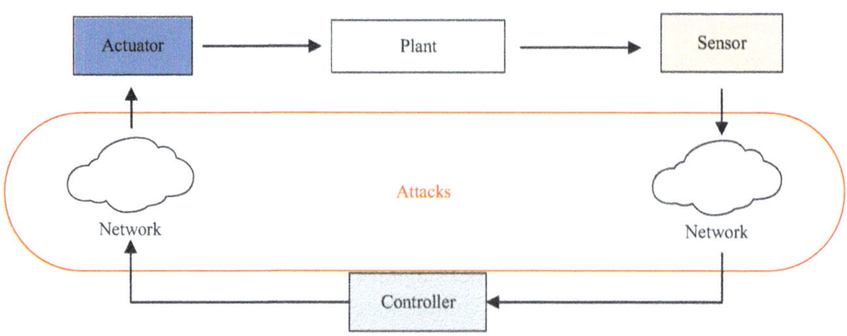

Fig. 1.6 Attacked NCS

[79] analyzed the problem of false data injection (FDI) attacks against the output tracking control of a network system. From the perspective of the attackers, stealth FDI attacks are recommended for feedback and forward channels to decrease the stability of the closed-loop system and avoid detection based on Kalman filters. Lu and Yang [67] investigated the control of event-triggered security-observer-based continuous-time network physical systems under actuator and sensor attacks. A novel security observer-based controller was proposed, and the actuator attack estimation was used to prevent actuator attacks.

Zhu and Martinez [170] proposed a variant of the receding-horizon control law for a discrete-time LTI system subjected to a replay attack and malevolently repeated messages sent from the operator to the actuator. Foroush and Martinez [17] developed an event-triggered control strategy to prevent energy-constrained periodic interference attacks. Gupta et al. [26] designed an optimal control law for intelligent jammers with finite actions. The attacker tried to determine the optimum attack strategy by maximizing a linear quadratic Gaussian cost function. Therefore, Zhang et al. [153] designed an optimal attack schedule to prevent attacks against NCSs.

Nearly all existing approaches are theoretical. Numerous studies have analyzed the usefulness of NCS in critical applications. To make matters worse, cyber-attacks are becoming more sophisticated. Again, we use the smart car as an example. Attacks on intelligent vehicles include FDI, deceive attack and channel control, data and information, unauthorized access, DoS, noise attack, privacy attack, tampering with data transmission, data integrity, illegal access to data transmissions, message replay history, identity theft, pretending to have legal status, and communication access. Therefore, it is urgent to develop effective network security defense strategies.

1.6 Conclusions

In this chapter, we reviewed the latest NCS research challenges. The analysis and design of research topics in NCS were discussed, including network latency, packet loss, packet clutter, interference, uncertainty, data quantification, event triggering, sensor or actuator failure, scheduling protocols, network security, and topology. Various mathematical models with different constraint conditions were described, and the corresponding control methods to solve these problems were illustrated. According to the reviewed literature, it can be concluded that NCSs have made significant progress over the past few decades, but there are several challenges that need to be considered in future work. We highlight some of them as follows.

(1) Control structure: A distributed grid and long-distance data transmission are typical in complex NCSs, such as smart grid, big data, cloud control, and remote control systems. Traditional control structures may not be able to cope with the new challenges. Therefore, it is necessary to develop a new control structure to address problems in newly emerging complex control systems.

(2) Engineering application: Significant achievements were made in theoretical NCS research in the past decades. However, only some applications were successfully implemented in a few industries. How to make full use of the theory in practical engineering applications, such as the smart grid, big data, cloud control, remote control systems, the Internet of Things, and space station robots, is a topic worth discussing.
(3) Control objects: most studies focused on simple nodes and simple systems, multiple sensors (or multiple nodes), multiple systems, and the coupling of various nodes or subsystems.
(4) Control algorithm: since a complex NCS is characterized by a large area and big data, network control techniques should be combined with computer algorithms, cloud storage, data mining, and wide-area measurements to develop some more effective control algorithms.
(5) Cyber security: In addition to the common challenges mentioned above, cyber security will be a major challenge in the future, especially attacks on and the defense of networks to ensure the quality of service (QoS). However, cyber security is still in its infancy.

References

1. Amin S, Cárdenas AA and Sastry SS (2009) Safe and secure networked control systems under Denial-of-Service attacks. In: Proceedings of the 12th International Conference on Hybrid Systems: Computation and Control. San Francisco, USA, 13–15 April 2009, pp. 31–45.
2. Amin S, Schwartz GA and Sastry SS (2013) Security of interdependent and identical networked control systems. Automatica 49(1): 186–192.
3. Anta A and Tabuada P (2010) To sample or not to sample: Self-triggered control for nonlinear systems. IEEE Transactions on Automatic Control 55(9): 2030–2042.
4. An L and Yang GH (2018) Improved adaptive resilient control against sensor and actuator attacks. Information Sciences 423: 145–156.
5. Baños A, Perez F and Cervera J (2014) Network-based reset control systems with time-varying delays. IEEE Transactions on Industrial Informatics 10(1): 514-522.
6. Branicky MS, Phillips SM and Zhang W (2002) Scheduling and feedback co-design for networked control systems. In: Proceedings of the 41st IEEE Conference on Decision and Control, Las Vegas, NV, USA, 10–13 December 2002, pp. 1211–1217.
7. Chen C, Zhang K, Yuan K, Zhu L and Qian M (2017) Novel detection scheme design considering cyber-attacks on load frequency control. IEEE Transactions on Industrial Informatics. Epub ahead of print 23 October 2017. https://doi.org/10.1109/TII.2017.2765313.
8. Chen HF, Gao JF, Shi T and Lu RQ (2016) H∞ control for networked control systems with time delay, data packet dropout and disorder. Neurocomputing 179: 211–218.
9. Chen Z, Zhang B, Li H and Yu J (2016) Tracking control for polynomial fuzzy networked systems with repeated scalar nonlinearities. Neurocomputing 171: 185-193.
10. Cho HC and Fadali MS (2011) Nonlinear network-induced time delay systems with stochastic learning. IEEE Transactions on Control Systems Technology 19(4): 843- 851.
11. Cloosterman MBG, Wouw NVD, Heemels WPMH and Nijmeijer H (2009) Stability of networked control systems with uncertain time-varying delays. IEEE Transactions on Automatic Control 54(7): 1575-1580.

12. Cuenca Á, Salt J, Sala A and Pizá R (2011) A delay-dependent dual-rate PID controller over an Ethernet network. IEEE Transactions on Industrial Informatics 7(1): 18-29.
13. Dai SL, Lin H and Ge SS (2010) Scheduling-and-control codesign for a collection of networked control systems with uncertain delays. IEEE Transactions on Control Systems Technology 18(1): 66-78.
14. Donkers MCF, Heemelsa WPMH, Bernardini D, Bemporadc A and Shneer V (2012) Stability analysis of stochastic networked control systems. Automatica 48(5): 917-925.
15. Du SL, Sun XM and Wang W (2014) Guaranteed cost control for uncertain networked control systems with predictive scheme. IEEE Transactions on Automation Science and Engineering 11(3): 740-748.
16. Feng Z, Lam J (2012) Robust reliable dissipative filtering for discrete delays singular systems. Signal process 92(12): 3010–3025.
17. Foroush HS and Martínez S (2012) On event-triggered control of linear systems under periodic Denial-of-Service jamming attacks. In: 2012 IEEE 51st Annual Conference on Decision and Control (CDC), Maui, HI, USA, 10–13 Dec. 2012, pp. 2551–2556.
18. Franzè G, Tedesco F and Famularo D (2015) Model predictive control for constrained networked systems subject to data losses. Automatica 54: 272-278.
19. Freirich D and Fridman E (2016) Decentralized networked control of systems with local networks: A time-delay approach. Automatica 69: 201-209.
20. Gaid MEMB, Çela A and Hamam Y (2006) Optimal integrated control and scheduling of networked control systems with communication constraints: Application to a car suspension system. IEEE Transactions on Control Systems Technology 14(4): 776–787.
21. Gao HJ, Chen TW and Lam J (2008) A new delay system approach to network-based control. Automatica 44(1): 39-52.
22. Gao M, Sheng L, Liu Y and Zhu Z (2016) Observer-based H∞ fuzzy control for nonlinear stochastic systems with multiplicative noise and successive packet dropouts. Neurocomputing 173: 2001-2008.
23. Ge X and Han QL (2017) Consensus of multiagent systems subject to partially accessible and overlapping Markovian network topologies. IEEE Transactions on Cybernetics 47(8): 1807-1819.
24. Geromel JC, Gabriel GM (2015) Optimal H2 state feedback sampled-data control design of Markov jump linear systems. Automatica 54: 182-188.
25. Guo J, Zhao Y, Sun CY and Yu Y (2015) Recursive identification of FIR systems with binary-valued outputs and communication channels. Automatica 60: 165-172.
26. Gupta A, Langbort C and Basar T (2010) Optimal control in the presence of an intelligent jammer with limited actions. In: 2010 49th IEEE Conference on Decision and Control (CDC), Atlanta, GA, USA, 15–17 Dec. 2010, pp. 1096–1101.
27. Gupta RA and Chow MY (2008) Performance assessment and compensation for secure networked control systems. In: Process of Industrial Electronics Society, Orlando, FL, USA, 10–13 November 2008, pp. 2929–2934.
28. Gupta RA and Chow MY (2010) Networked control system: overview and research trends. IEEE Transactions on Industrial Electronics 57(7): 2527- 2535.
29. Gupta RA, Chow MY, Agarwal AK and W Wang (2007) Information security with real-time operation: Performance assessment for next generation wireless distributed networked-control-systems. In: Process of Industrial Electronics Society, Taipei, Taiwan, 5–8 November 2007, pp. 420–425.
30. Heemels WPMH, Teel AR, Wouw NVD, Nesic D (2010) Networked control systems with communication constraints: tradeoffs between transmission intervals, delays and performance. IEEE Transactions on Automatic Control 55(8): 1781–1796.
31. Hu WS, Liu GP and Rees D (2007) Event-driven networked predictive control. IEEE Transactions on industrial electronics 54(3): 1603-1613.
32. Kang X and Ishii H (2015) Coarsest quantization for networked control of uncertain linear systems. Automatica 51: 1-8.

33. Kar S and Moura JM (2010) Distributed consensus algorithms in sensor networks: quantized data and random link failures. IEEE Transactions on Signal Processing 58(3): 1383–1400.
34. Khanesar MA, Kaynak O, Yin S and Gao H (2015) Adaptive indirect fuzzy sliding mode controller for networked control systems subject to time-varying network-induced time delay. IEEE Transactions on Fuzzy Systems 23(1): 205-214.
35. Khatib EJ, Barco R, Muñoz P, et al. (2016) Self-healing in mobile networks with big data. IEEE Communications Magazine 54(1): 114-120.
36. Kima JH and Park HB (1999) H∞state feedback control for generalized continuous/discrete time-delay system. Automatica 35(8): 1443-1451.
37. Kwon C and Hwang I (2017) Reachability analysis for safety assurance of cyber-physical systems against cyber-attacks. IEEE Transactions on Automatic Control. Epub ahead of print 10 October 2017. https://doi.org/10.1109/TAC.2017.2761762.
38. Li A, Yi S and Wang X (2016) New reliable H∞ filter design for networked control systems with external disturbances and randomly occurring sensor faults. Neurocomputing 185: 21-27.
39. Li H, Chow MY and Sun Z (2009) Optimal stabilizing gain selection for networked control systems with time delays and packet losses. IEEE Transactions on Control Systems Technology 17(5): 1154–1162.
40. Li H, Yang H, Sun F and Xia Y (2014) Sliding-mode predictive control of networked control systems under a multiple-packet transmission policy. IEEE Transactions on Industrial Electronics 61(11): 6234-6243.
41. Li H, Wu C, Shi P and Gao Y (2015) Control of nonlinear networked systems with packet dropouts: interval type-2 fuzzy model-based approach. IEEE Transactions on Cybernetics 45(11): 2378-2389.
42. Li J, Er MJ, Yu H (2016) Sampling and control strategy: networked control systems subject to packet disordering. IET Control Theory & Applications 10(6): 674–683.
43. Li M and Chen Y (2017a) Sliding mode predictive tracking control for dc permanent magnet motor in NCSs with random delay and packet dropouts. International Journal of Control, Automation and Systems 15(6): 2723-2732.
44. Li M and Chen Y (2017b) Robust tracking control of networked control systems with communication constraints and external disturbance. IEEE Transactions on Industrial Electronics 64(5): 4037-4047.
45. Li M and Chen Y (2017c) A wide-area dynamic damping controller based on robust H∞ control for wide-area power systems with random delay and packet dropout. IEEE Transactions on Power Systems. Epub ahead of print 13 December 2017. https://doi.org/10.1109/TPWRS.2017.2782792.
46. Li M and Chen Y (2018a) Wide-area stabiliser on sliding mode control for cross-area power systems with random delay and packet dropouts. Asian Journal of Control 20(6): 1–13.
47. Li M and Chen Y (2018b) Robust adaptive sliding mode control for switched networked control systems with disturbance and faults. IEEE Transactions on Industrial Informatics. Epub ahead of print 23 February 2018. https://doi.org/10.1109/TII.2018.2808921.
48. Li M, Chen Y, Zhang XB and Xian YP (2016) A predictive control method for networked control systems with random delay and packet dropouts. In: Proceedings of the 35th Chinese Control Conference, Chengdu, China, 27–29 July 2016, pp. 4372–4377.
49. Li XY, Sun SL (2015) H∞ filtering for networked linear systems with multiple packet dropouts and random delays. Digital Signal Processing 46: 59-67.
50. Li Y, Shi L, Cheng P, Chen J and Quevedo DE (2013) Jamming attack on cyber-physical systems: A game-theoretic approach. In: 2013 IEEE 3rd Annual International Conference on Control and Intelligent Systems (CYBER), Nanjing, China, 26–29 May 2013, pp. 252–257.
51. Liberzon D (2006) Quantization, time delays, and nonlinear stabilization. IEEE Transactions on Automatic Control 51(7): 1190–1195.
52. Lin H, Su H, Shu Z, Wu ZG and Xu Y (2016) Optimal estimation in UDP-like networked control systems with intermittent inputs: stability analysis and suboptimal filter design. IEEE Transactions on Automatic Control 61(7): 1794–1809.

53. Lin Z, Han T, Zheng R and Yu C (2017) Distributed localization with mixed measurements under switching topologies. Automatica 76: 251–257.
54. Liu A, Zhang WA, Yu L, Liu S, Chen MZQ (2015) New results on stabilization of networked control systems with packet disordering. Automatica 52: 255-259.
55. Liu B and Xia Y (2011) Fault detection and compensation for linear systems over networks with random delays and clock asynchronism. IEEE Transactions on Industrial Electronics 58(9): 4396-4406.
56. Liu GP, Xia YQ, Rees D and Hu WS (2007) Design and stability criteria of networked predictive control systems with random network delay in the feedback channel. IEEE Transactions on Systems, Man, and Cybernetics—Part C: Applications and Reviews 37(2): 173–184.
57. Liu J and Yue D (2013) Event-based fault detection for networked systems with communication delay and nonlinear perturbation. Journal of the Franklin Institute 350(9): 2791–2807.
58. Liu J, Fei S, Tian E and Gu Z (2015) Co-design of event generator and filtering for a class of T-S fuzzy systems with stochastic sensor faults. Fuzzy Sets and Systems: 273: 124–140.
59. Liu K, Fridman E and Hetel L (2014) Networked control systems in the presence of scheduling protocols and communication delays. Siam Journal on Control & Optimization 53(4): 2808-2817.
60. Liu K, Fridmanb E and Johansson KH (2015a) Dynamic quantization of uncertain linear networked control systems. Automatica 59: 248-255.
61. Liu K, Fridman E and Johansson KH (2015b) Networked control with stochastic scheduling. IEEE Transactions on Automatic Control 60(11): 3071-3076.
62. Liu X, Yu X, Ma G and Xi H (2016) On sliding mode control for networked control systems with semi-Markovian switching and random sensor delays. Information Sciences 337: 44-58.
63. Liu Y, Wang Z and Wang W (2011) Reliable $H\infty$ filtering for discrete time-delay systems with randomly occurred nonlinearities via delay-partitioning method. Signal Processing 91(4): 713–727.
64. Liu YY, Slotine JJ and Barabsi AL (2011) Controllability of complex networks. Nature 473(7346): 167-173.
65. Lješnjanin M, Nešić D, Quevedo DE (2016) Robust stability of a class of networked control systems. Automatica 73: 117-124.
66. Lunze J (2017) Structural properties of networked systems with random communication links. Automatica 80: 300–304.
67. Lu A and Yang G (2017) Event-triggered secure observer-based control for cyber-physical systems under adversarial attacks. Information Sciences 420: 96–109.
68. Lu R, Cheng H and Bai J (2015) Fuzzy-model-based quantized guaranteed cost control of nonlinear networked systems. IEEE Transactions on Fuzzy Systems 23(3): 567-575.
69. Lu R, Xu Y and Zhang R (2016) A new design of model predictive tracking control for networked control system under random packet loss and uncertainties. IEEE Transactions on Industrial Electronics 63(11): 6999-7007.
70. Matiakis T, Hirche S and Buss M (2009) Control of networked systems using the scattering transformation. IEEE Transactions on Control Systems Technology 17(1): 60-67.
71. Mo Y, Chabukswar R and Sinopoli B (2014) Detecting integrity attacks on SCADA systems. IEEE Transactions on Control Systems Technology 22(4): 1396–1407.
72. Moon J and Basar T (2015) Minimax control over unreliable communication channels. Automatica 59:182-193.
73. Donkers MCF and Heemelsa WPMH (2012) Output-based event-triggered control with guaranteed $L\infty$-gain and improved and decentralized event-triggering. IEEE Transactions on Automatic Control .57(6):1362–1376.
74. Nesic D and Liberzon D (2009) A unified framework for design and analysis of networked and quantized control systems. IEEE Transactions on Automatic Control 54(4): 732–747.
75. Nowzari C and Cortés J (2012) Self-triggered coordination of robotic networks for optimal deployment. Automatica 48(6): 1077–1087.
76. Pang ZH, Liu GP, Zhou DH and Chen M (2014) Output tracking control for networked systems: a model-based prediction approach. IEEE Transactions on Industrial Electronics 61(9): 4867-4877.

77. Pang ZH, Liu GP, Zhou DH, Sun DH (2015) Data-based predictive control for networked non-linear systems with two-channel packet dropouts. IET Control Theory and Applications 9(7): 1154-1161.
78. Pang ZH, Liu G, Zhou DH and Sun DH (2016a) Data-based predictive control for networked nonlinear systems with network-induced delay and packet dropout. IEEE Transactions on Industrial Electronics 63(2): 1249-1257.
79. Pang ZH, Liu GP, Zhou D, Hou F and Sun D (2016b) Two-channel false data injection attacks against output tracking control of networked systems. IEEE Transactions on Industrial Electronics 63(5): 3242-3251.
80. Pasqualetti F, Dörfler F and Bullo F (2013) Attack detection and identification in cyber-physical systems. IEEE Transactions on Automatic Control 58(11): 2715- 2729.
81. Peng C and Han QL (2013) A novel event-triggered transmission scheme and L2 control co-design for sampled-data control systems. IEEE Transactions on Automatic Control 58(10): 2620–2626.
82. Peng C and Han QL (2016) On designing a novel self-triggered sampling scheme for networked control systems with data losses and communication delays. IEEE Transactions on Industrial Electronics 63(2): 1239-1248.
83. Probst A, Magana ME and Sawodny O (2010) Using a Kalman filter and a Pade´approximation to estimate random time delays in a networked feedback control system. IET Control Theory and Applications 4(11): 2263–2272.
84. Psillakis HE (2017) Consensus in networks of agents with unknown high-frequency gain signs and switching topology. IEEE Transactions on Automatic Control 62(8): 3993- 3998.
85. Qin J, Yu C and Gao H (2015) Collective behavior for group of generic linear agents interacting under arbitrary network topology. IEEE Transactions on Control of Network Systems 2(3): 288-297.
86. Qiu J, Gao H and Ding SX (2016) Recent advances on fuzzy-model-based nonlinear networked control systems: a survey. IEEE Transactions on Industrial Electronics 63(2): 1207-1217.
87. Qiu L, Shi Y, Yao F, Xu G and Xu B (2015) Network-based robust H2/H∞ control forlinear systems with two-channel random packet dropouts and time delays. IEEE Transactions on Cybernetics 45(8): 1450-1462.
88. Qiu L, Yao F, Xu G, Li S and Xu B (2016) Output feedback guaranteed cost control for networked control systems with random packet dropouts and time delays in forward and feedback communication links. IEEE Transactions on Automation Science and Engineering 13(1): 284- 295.
89. Qu F, Hu B, Guan Z, Wu Y, He D and Zheng D (2016) Quantized stabilization of wireless networked control systems with packet losses. ISA Transactions 64: 92-97.
90. Quevedo DE and Jurado I (2014) Stability of sequence-based control with random delays and dropouts. IEEE Transactions on Automatic Control 59(5): 1296-1302.
91. Rakkiyappan R, Sakthivel N, Cao J (2015) Stochastic sampled-data control for synchronization of complex dynamical networks with control packet loss and additive time-varying delays. Neural networks 66:46–63.
92. Repele L, Muradore R, Quaglia D and Fiorini P (2014) Improving performance of networked control systems by using adaptive buffering. IEEE Transactions on Industrial Electronics 61(9): 4847-4856.
93. Ruths J and Ruths D (2014) Control profiles of complex networks. Science 343(6177): 1373–1376.
94. Shen D, Zhang C and Xu Y (2017) Two updating schemes of iterative learning control for networked control systems with random data dropouts. Information Sciences 381: 352–370.
95. Shen QK, Jiang B, Shi P and Zhao J (2014) Cooperative adaptive fuzzy tracking control for networked unknown nonlinear multi-agent systems with time-varying actuator faults. IEEE Transactions on Fuzzy Systems 22(3): 494 -504.
96. Shi Y, Huang J and Yu B (2013) Robust tracking control of networked control systems: application to a networked dc motor. IEEE Transactions on Industrial Electronics 60(12): 5864-5874.

97. Song Y, Wang J, Shi Y, Li C (2013) H∞ control of networked control systems with time delay and packet disordering. Journal of the Franklin Institute 350(6): 1596–1616.
98. Su X, Wu L and Shi P (2013) Sensor networks with random link failures: distributed filtering for T–S fuzzy systems. IEEE Transactions on Industrial informatics 9(3): 1739–1750.
99. Sun XM, Liu KZ, Wen C, Wang W (2016) Predictive control of nonlinear continuous networked control systems with large time-varying transmission delays and transmission protocols. Automatica, vol. 64, pp. 76-85.
100. Tabbara M and Nesic D (2008) Input-output stability of networked control systems with stochastic protocols and channels. IEEE Transactions on Automatic Control 53(5): 1160–1175.
101. Tan C, Li L and Zhang H (2015) Stabilization of networked control systems with both network-induced delay and packet dropout. Automatica 59: 194-199.
102. Tan C and Liu GP (2013) Consensus of discrete-time linear networked multi-agent systems with communication delays. IEEE Transactions on Automatic Control 58(11): 2962-2968.
103. Tang B, Peng S and Zhang Y (2016) Fuzzy-model-based robust H∞ design of nonlinear packetized networked control systems. IEEE Transactions on Fuzzy Systems 24(3): 544-557.
104. Tao J, Lu RQ, Su HY, Shi P and Wu ZG (2017) synchronous filtering of nonlinear Markov jump systems with randomly occurred quantization via T-S fuzzy models. IEEE Transactions on Fuzzy Systems. Epub ahead of print 20 September 2017. https://doi.org/10.1109/TFUZZ.2017.2754999.
105. Ugrinovskii V and Fridman E (2014) A Round-Robin type protocol for distributed estimation with H∞ consensus. Systems & Control Letters 69: 103-110.
106. Walsh GC, Ye H and Bushnell LG (2002) Stability analysis of networked control systems. IEEE Transactions on Control Systems Technology 10(3): 438–446.
107. Wang GQ, Liu XQ and Shi P (2017) Observer-based fuzzy adaptive output-feedback control of stochastic nonlinear multiple time-delay systems. IEEE Transactions on Cybernetics 47(9): 2568-2578.
108. Wang R, Liu GP, Wang W, Rees D and Zhao YB (2010a) Control for networked predictive control systems based on the switched Lyapunov function method. IEEE Transactions on Industrial Electronics 57(10): 3565- 3571.
109. Wang R, Liu G, Wang W, Rees D and Zhao YB (2010b) Guaranteed cost control for networked control systems based on an improved predictive control method. IEEE Transactions on Control Systems Technology 18(5): 1226-1232.
110. Wang R, Liu GP, Wang W, Rees D and Zhao YB (2010c) H∞ control for networked predictive control systems based on the switched Lyapunov function method. IEEE Transactions on Industrial Electronics 57(10): 3565- 3571.
111. Wang S, Fang H and Tian X (2015) Recursive estimation for nonlinear stochastic systems with multi-step transmission delays, multiple packet dropouts and correlated noises. Signal Processing 115: 164-175.
112. Wang X, Kharisov E and Hovakimyan N (2015) Real-time L1 adaptive control for uncertain networked control systems. IEEE Transactions on Automatic Control 60(9): 2500-2505.
113. Wang X and Lemmon MD (2009) Self-triggered feedback control systems with finite-gain l2 stability. IEEE Transactions on Automatic Control 54(3): 452–467.
114. Wei YL, Qiu JB, Shi P and Wu LG (2017) A piecewise-Markovian Lyapunov approach to reliable output feedback control for fuzzy-affine systems with time-varying delays and actuator faults. IEEE Transactions on Cybernetics. Epub ahead of print 4 December 2017. https://doi.org/10.1109/TCYB.2017.2749239.
115. Wu X, Zhao X, Lü J, Tang L and Lu JA (2016) Identifying topologies of complex dynamical networks with stochastic perturbations. IEEE Transactions on Control of Network Systems 3(4): 379-389.
116. Wu ZG, Xu ZW, Shi P, Chen MZQ and Su HY (2018) Non-fragile state estimation of quantized complex networks with switching topologies. IEEE Transactions on Neural Networks and Learning Systems. Epub ahead of print 1 February 2018. https://doi.org/10.1109/TNNLS.2018.2790982.

117. Wang Y, Ding SX, Ye H, and Wang G (2008) A new fault detection scheme for networked control systems subject to uncertain time-varying delay. IEEE Transactions on Signal Processing 56(10): 5258-5268.
118. Wang Y, Xia M, Gupta V and Antsaklis P (2015) On feedback passivity of discrete-time nonlinear networked control systemswith packet drops. IEEE Transactions on Automatic Control 60(9): 2434–2439.
119. Wang YL and Han QL (2014) Modelling and observer-based H∞ controller design for networked control systems. IET Control Theory and Applications 8(15): 1478–1486.
120. Wang YL, Shi P, Lim CC and Liu Y (2016) Event-triggered fault detection filter design for a continuous-time networked control system. IEEE Transactions on Cybernetics 46(12): 3414-3426.
121. Wen G, Yu W, Hu G, Cao J and Yu X (2015) Pinning synchronization of directed networks with switching topologies: a multiple Lyapunov functions approach. IEEE Transactions on Neural Networks and Learning Systems 26(12): 3239-3249.
122. Wu D, Sun XM, Wang W, Shi P (2014) Robust predictive control for networked control and application to DC-motor control. IET Control Theory & Applications 8(14): 1312-1320.
123. Wu D, Sun XM, Wen C and Wang W (2016) Redesigned predictive event-triggered controller for networked control system with delays. IEEE Transactions on Cybernetics 46(10): 2195-2206.
124. Wu H, Lou L, Chen CC, Hirche S and Kühnlenz K (2013) Cloud-based networked visual servo control. IEEE Transactions on Industrial Electronics 60(2): 554-566.
125. Wu W, Reimann S, Görges D and Liu S (2015) Suboptimal event-triggered control for time-delayed linear systems. IEEE Transactions on Automatic Control 60(5): 1386-1391.
126. Xia YQ (2015) Cloud control systems. IEEE/CAA Journal of Automatica Sinica 2(2):134-142.
127. Xia YQ, Gao YL, Yan LP and Fu MY (2015) Recent progress in networked control systems — A survey. International Journal of Automation and Computing 12(4): 343-367.
128. Xu Y, Lu RQ, Shi P, Tao J and Xie SL (2018) Robust estimation for neural networks with randomly occurring distributed delays and Markovian jump coupling. IEEE Transactions on Neural Networks and Learning Systems 29(4): 845-855.
129. Xu Y, Lu RQ, Zhou KX and Li Z (2016) Nonfragile asynchronous control for fuzzy Markov jump systems with packet dropouts. Neurocomputing 175: 443-449.
130. Xu ZW, Su HY, Shi P, Lu RQ and Wu ZG (2017) Reachable set estimation for Markovian jump neural networks with time-varying delays. IEEE Transactions on Cybernetics 47(10): 3208 -3217.
131. Xua H, Jagannathan S, Lewis FL (2012) Stochastic optimal control of unknown linear networked control system in the presence of random delays and packet losses. Automatica 48(6): 1017-1030.
132. Yan H, Wang T, Zhang H and Shi H (2015) Event-triggered H∞ control for uncertain networked T–S fuzzy systems with time delay. Neurocomputing 157: 273–279.
133. Yan H, Zhang H, Shi H and Meng QH (2011) Improved robust H∞ performance analysis for networked control systems with two additive time-varying delays. In: 9th World Congress on Intelligent Control and Automation (WCICA), Taipei, Taiwan, 21–25 June 2011, pp. 827–832.
134. Yang R, Liu GP, Shi P, Thomas C and Basin M (2014) Predictive output feedback control for networked control systems. IEEE Transactions on Industrial Electronics 61(1): 512- 520.
135. Yin X, Zhang L, Zhu Y, Wang C and Li Z (2016) Robust control of networked systems with variable communication capabilities and application to a semi-active suspension system. IEEE/ASME Transactions on Mechatronics 21(4): 2097-2107.
136. Yin S, Gao H, Qiu J and Kaynak O (2017) Descriptor reduced-order sliding mode observers design for switched systems with sensor and actuator faults. Automatica 76: 282-292.
137. Yu J, Nan L, Tang X and Wang P (2015) Model predictive control of non-linear systems over networks with data quantization and packet loss. ISA Transactions 59: 1-9.
138. Yue D, Han QL and Lam J (2005) Network-based robust H∞ control of systems with uncertainty. Automatica 41(6): 999-1007.

139. Yuan Y, Yuan H, Guo L, Yang H and Sun S (2016) Resilient control of networked control system under dos attacks: a unified game approach. IEEE Transactions on Industrial Informatics 12(5): 1786-1794.
140. Yue D, Tian E and Han QL (2013) A delay system method for designing event-triggered controllers of networked control systems. IEEE Transactions on Automatic Control 58(2): 475–481.
141. Zhai DH and Xia Y (2016) Adaptive control for teleoperation system with varying time-delays and input saturation constraints. IEEE Transactions on Industrial Electronics 63(11): 6921-6929.
142. Zhan XS, Sun XX, Wu J, Han T (2016) Optimal modified tracking performance for networked control systems with QoS constraint. ISA Transactions 65: 109-115.
143. Zhan XS, Wu J, Jiang T and Jiang XW (2015) Optimal performance of networked control systems under the packet dropouts and channel noise. ISA Transactions 58: 214-221.
144. Zhang D, Nguang SK, Srinivasan D and Yu L (2018) Distributed filtering for discrete-time T–S fuzzy systems with incomplete measurements. IEEE Transactions on Fuzzy Systems 26(3): 1459-1471.
145. Zhang D, Nguang SK and Yu L (2017) Distributed control of large-scale networked control systems with communication constraints and topology switching. IEEE Transactions on Systems, Man, and Cybernetics: Systems 47(7): 1746-1757.
146. Zhang D, Wang QG, Srinivasan D, Li HY and Yu L (2018) Asynchronous state estimation for discrete-time switched complex networks with communication constraints. IEEE Transactions on Neural Networks and Learning Systems 29(5): 1732-1746.
147. Zhang D, Shi P, Wang QG and Yu L (2017) Analysis and synthesis of networked control systems: A survey of recent advances and challenges. ISA Transactions 66: 376-392.
148. Zhang D, Shi P, Zhang WA and Yu L (2017) Energy-efficient distributed filtering in sensor networks: A unified switched system approach. IEEE Transactions on Cybernetics, 47(7): 1618-1629.
149. Zhang D, Yu L, Wang QG and Ong C (2012) Estimator design for discrete-time switched neural networks with asynchronous switching and time-varying delay. IEEE Transactions on Neural Networks and Learning Systems 23(5): 827-834.
150. Zhang H, Cheng P, Shi L and Chen J (2016) Optimal DoS attack scheduling in wireless networked control system. IEEE Transactions on Control Systems Technology 24(3): 843-852.
151. Zhang H, Shi Y and Liu M (2013) H∞ step tracking control for networked discrete-time nonlinear systems with integral and predictive actions. IEEE Transactions on Industrial Informatics 9(1): 337-345.
152. Zhang H, Yang D and Chai T (2007) Guaranteed cost networked control for T–S fuzzy systems with time delays. IEEE Transactions on Systems, Man, and Cybernetics—Part C: Applications and Reviews 37(2): 160–172.
153. Zhang H, Yue D, Yin X and Chen J (2016) Adaptive model-based event-triggered control of networked control system with external disturbance. IET Control Theory & Applications 10(15): 1956-1962.
154. Zhang J, Zhang H, Luo Y, Feng T (2014) Model-free optimal control design for a class of linear discrete-time systems with multiple delays using adaptive dynamic programming. Neurocomputing 135: 163-170.
155. Zhang JH, Xia YQ and Shi P (2013) Design and stability analysis of networked predictive control systems. IEEE Transactions on Control Systems Technology 21(4): 1495-1501.
156. Zhang L, Shi Y, Chen T and Huang B (2005) A new method for stabilization of networked control systems with random delays. IEEE Transactions on Automatic Control 50(8): 633-637.
157. Zhang L and Varsakelis DH (2006) Communication and control co-design for networked control systems. Automatica 42(6): 953–958.
158. Zhang LX, Gao HJ and Kaynak O (2013) Network-induced constraints in networked control systems—A survey. IEEE Transactions on Industrial Informatics 9(1): 403- 416.

159. Zhang LX, Ning ZP and Shi P (2015) Input-output approach to control for fuzzy Markov jump systems with time-varying delays and uncertain packet dropout rate. IEEE Transactions on Cybernetics 45(11): 2449 – 2460.
160. Zhang M, Shi P, Ma LH, Cai JP and Su HY (2018) Quantized feedback control of fuzzy Markov jump systems. IEEE Transactions on Cybernetics. DOI: https://doi.org/10.1109/TCYB.2018.2842434.
161. Zhang M, Shi P, Liu Z, Ma LH, Cai JP and Su HY (2018) H∞ filtering for discrete-time switched fuzzy systems with randomly occurring time-varying delay and packet dropouts. Signal Processing 143: 320-327.
162. Zhang W and Yu L (2007) Output feedback stabilization of networked control systems with packet dropouts. IEEE Transactions on Automatic Control 52(9): 1705-1710.
163. Zhang WA and Yu L (2008) Modelling and control of networked control systems with both network-induced delay and packet-dropout. Automatica 44(12): 3206-3210.
164. Zhang WA, Yu L and Feng G (2011) Optimal linear estimation for networked systems with communication constraints. Automatica 47(9): 1992–2000.
165. Zhang XM, Han QL and Yu XH (2016) Survey on recent advances in networked control systems. IEEE Transactions on Industrial Informatics 12(5): 1740-1752.
166. Zhao YB, Kim J, Liu GP, Rees D (2012) Compensation and stochastic modeling of discrete-time networked control system with data packet disorder. International Journal of Control, Automation and Systems 10(5): 1055–1063.
167. Zhao YB, Liu GB and Rees D (2008) Integrated predictive control and scheduling co-design for networked control systems. IET Control Theory & Applications 2(1), pp. 7–15.
168. Zhou Z, Lin T, Thulasiraman K and Xue G (2017) Novel survivable logical topology routing by logical protecting spanning trees in IP-Over-WDM networks. IEEE/ACM Transactions on Networking 25(3): 1673-1685.
169. Zhu Y, Zhang L, Ning Z, Zhu Z, Shammakh W and Hayat T (2015) H∞ state estimation for discrete-time switching neural networks with persistent dwell-time switching regularities. Neurocomputing 165: 414–422.
170. Zhu M and Martínez S (2014) On the performance analysis of resilient networked control systems under replay attacks. IEEE Transactions on Automatic Control 59(3): 804–808.
171. Zou A, Wang Z and Gao H (2016) Observer-based H∞ control of networked systems with stochastic communication protocol: The finite-horizon case. Automatica 63: 366-373.

Chapter 2
Adaptive Sliding-Mode Tracking Control for Networked Systems with Network-Induced Disturbance

The intelligent vehicle is a typical networked control system. This chapter focuses on the tracking control of intelligent vehicles with network-induced interference. An adaptive sliding mode tracking control method is presented [1]. First, an adaptive sliding mode surface based on the error of the tracking output is proposed. The estimated value of the external disturbance is used as an adaptive factor to mitigate the external disturbance. Second, an adaptive sliding mode controller is designed, and the convergence of the sliding mode dynamics is demonstrated. Third, we consider the network-induced disturbance affecting the sensor output. A novel disturbance observer is proposed to compensate for the network-induced disturbance. Finally, the error of the tracking output is discussed, and the effectiveness of the designed observer and proposed adaptive sliding mode tracking control approach is demonstrated using a simulation example.

2.1 Introduction

Vehicles have improved peoples' lives. Due to technological developments in communication and vehicle design in recent years, intelligent vehicles have attracted the attention of many researchers [2, 3]. An intelligent vehicle is a typical networked control system; the vehicle network is used to transmit data and control the intelligent vehicle. This system has the advantages of a networked control system, such as high reliability, easy extensibility, and resource sharing. However, vehicles on the road are inevitably affected by external disturbances. In addition, a wired/wireless network also causes many disturbances, such as sensor failure, measurement noise, and hacking attacks. Thus, existing algorithms may result in low control performance or tracking failure. Therefore, the prevention of network-induced disturbances is a hot topic in current research.

Sliding mode control (SMC) is a robust control method with a variable structure and is insensitive to system disturbances [4, 5]. SMC provides good control

© The Author(s), under exclusive license to Springer Nature Singapore Pte Ltd. 2022

M. Li et al., *Tracking Control of Networked Systems via Sliding-Mode*,
https://doi.org/10.1007/978-981-16-6514-1_2

performance even when the system is affected by external interference. Therefore, the method is widely used in various industrial control fields. In [6], an extended state observer based on SMC was developed to suppress external disturbances in a grid-connected power converter system. In [7], an optimized SMC strategy was investigated to mitigate undesired voltage disturbances in single-phase dynamic voltage restorers. In recent years, many continuous industrial control systems have been transformed into discrete-time systems. As a result, discrete sliding mode control (DSMC) has received significant attention. In [8], a disturbance compensator was combined with SMC to minimize disturbances in a discrete-time velocity servo system. In [9], an adaptive robust SCM approach for switched networked control systems with external disturbance and actuator or sensor faults was proposed. Although it was reported in some studies that SMC provided good performance, some problems remain to be solved; for example, a critical problem is jitter.

Many control algorithms have been proposed to alleviate the influence of jitter [10, 11]. High-order SMC has attracted widespread attention. It alleviates jitter by ensuring that the sliding mode function and its derivative approach zero. In [12], a second-order discrete-time terminal SCM strategy was proposed to track the motion of a piezoelectric nanopositioning system. Castaneda et al. [13] developed a second-order SMC strategy for a helicopter with three degrees of freedom and external perturbations to achieve trajectory tracking control. In recent years, several algorithms have been proposed to suppress jitter, such as adaptive control and reaching law control [14, 15].

Delay and data packet loss are often two key factors affecting the system control performance in networked control systems. They often affect the stability of the system and reduce its control performance [16, 17]. Many control approaches have been reported to mitigate these two issues. Examples include robust control [18], the switch system method [19], and the optimal control approach [20]. However, in these methods, the control system only passively accepts delay and data packet loss. A networked predictive control algorithm was proposed in [21] to overcome this problem. This method predicts the control input of the system for a period in the future based on the data received by the current control terminal. In this way, the method can effectively suppress the effects of the delay and data packet loss. Subsequently, many improved network predictive control (NPC) algorithms were reported [22, 23]. However, most NPC algorithms predict the control input of the system based on the delay state or measurement output, i.e., they compensate for the delay of the feedforward channel but cannot adequately handle the delay of the feedback channel. In addition, the delay and data packet loss in a networked system can be regarded as interference. In [24, 25], delay and data packet loss were modeled as network-induced interference, and robust control strategies were designed to suppress the impact of the interference.

2.2 Problem Statement

2.2.1 Problem Formulation and Modeling

Figure 2.1 shows the control structure of an intelligent vehicle system that uses the adaptive tracking control method; the actuator and smart sensor are located on the plant side. The main function of the digital sensor is to collect data and transmit it to the control terminal. The actuator is responsible for receiving control input commands and driving the vehicle. The feedback channel, which connects the sensor end to the control end, and the feedforward channel, which connects the control end to the actuator end, are linked by a wired or wireless network. In a networked intelligent vehicle system, the tracking control is often affected by networked-induced interference, such as measurement noise, hacker attacks, and sensor faults.

The mathematical model of an intelligent vehicle can be described as follows:

$$\dot{\tilde{v}}(t) = \tilde{a} + \tilde{d}_1(t) \tag{2.1a}$$

$$\dot{\tilde{a}}(t) = \tilde{f}(\tilde{v}, \tilde{a}) + \tilde{g}(\tilde{v})\tilde{u} + \tilde{d}_2(t) \tag{2.1b}$$

$$\tilde{f}(\tilde{v}, \tilde{a}) = -\frac{1}{\tilde{\varsigma}}\left(\dot{\tilde{v}} + \frac{\tilde{\sigma}\tilde{\eta}\tilde{c}}{2\tilde{m}}\tilde{v}^2 + \frac{\tilde{\varphi}}{\tilde{m}}\right) - \frac{\tilde{\sigma}\tilde{\eta}\tilde{c}\tilde{v}\tilde{a}}{\tilde{m}} \tag{2.1c}$$

$$\tilde{g}(\tilde{v}) = \frac{1}{\tilde{\varsigma}\tilde{m}} \tag{2.1d}$$

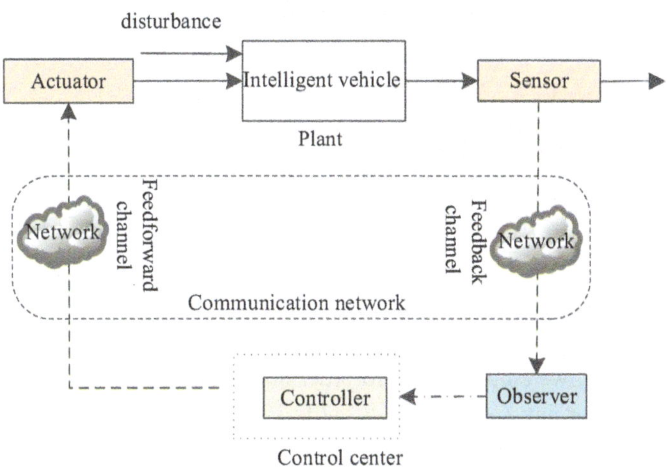

Fig. 2.1 Control structure of intelligent vehicle system

where $\tilde{v}$ and $\tilde{a}$ are the vehicle's velocity and acceleration, respectively, $\widehat{\varsigma}$, $\tilde{\sigma}$, $\tilde{m}$, $\tilde{c}$, $\tilde{\varphi}$, and $\tilde{\eta}$ stand for the engine time constant, the specific mass of the air, mass, drag coefficient, mechanical drag, and cross-sectional area, respectively, $\tilde{u}$ represents the engine input, $\tilde{d}_1(t)$ and $\tilde{d}_2(t)$ denote two disturbance terms, and both of them are bounded.

We adopt the following engine input in [13] to linearize the vehicle dynamics:

$$\tilde{u} = \tilde{\theta}\tilde{m} + \frac{1}{2}\tilde{\sigma}\tilde{\eta}\tilde{c}\tilde{v}^2 + \tilde{\varphi} + \tilde{\varsigma}\tilde{\sigma}\tilde{\eta}\tilde{c}\tilde{v}\tilde{a} \tag{2.2}$$

where $\tilde{\theta}$ is the input signal, which will be designed later.

By combining Eqs. (2.1a) and (2.2), the dynamic equation of acceleration in (2.1b) can be rewritten as:

$$\dot{\tilde{a}} = -\frac{1}{\tilde{\varsigma}}\tilde{a} + \frac{1}{\tilde{\varsigma}}\tilde{\theta} + \tilde{d}_2(t) - \frac{1}{\tilde{\varsigma}}\tilde{d}_1(t). \tag{2.3}$$

Remark 1 It can be seen that some characteristic parameters of the vehicle dynamics (such as the mass, air resistance, and mechanical drag) are excluded by introducing the engine input in Eq. (2.2). This approach not only linearizes the vehicle dynamics but also simplifies the dynamic equation of the system.

We define the new variable $\widehat{x}(t) = [\widehat{x}_1(t)\, \widehat{x}_2(t)]^T = [\tilde{v}\, \tilde{a}]^T, \widehat{d}(t) = [\tilde{d}_1(t)\, \tilde{d}_2(t)]^T$, and the control input $\widehat{u}(t) = \tilde{\theta}$. Then, according to Eqs. (2.1a) and (2.3), the state space equation of the intelligent vehicle can defined as:

$$\dot{\widehat{x}}(t) = \overline{A}\widehat{x}(t) + \overline{B}\widehat{u}(t) + \overline{D}\widehat{d}(t) \tag{2.4}$$

where $\overline{A} = \begin{pmatrix} 0 & 1 \\ 0 & -\frac{1}{\tilde{\varsigma}} \end{pmatrix}$, $\overline{B} = \begin{pmatrix} 0 \\ \frac{1}{\tilde{\varsigma}} \end{pmatrix}$, and $\overline{D} = \begin{pmatrix} 1 & 0 \\ -\frac{1}{\tilde{\varsigma}} & 1 \end{pmatrix}$.

The output equation is described as

$$\widehat{y}(t) = \tilde{v} = \overline{C}\widehat{x}(t) \tag{2.5}$$

where $\overline{C} = [\,1\;0\,]$ denotes the output gain matrix.

In this chapter, we employ the zero-order hold method to deal with the control input $\widehat{u}(t)$, i.e., for any $t \in [kT, (k+1)T]$, $\widehat{u}(t) = \widehat{u}(kT)$, where T denotes the sampling interval. Then, under the discrete-time control input $\widehat{u}(kT)$, the system Eqs. (2.4) and (2.5) can be rewritten in the following discrete form:

$$\begin{cases} \widehat{x}(k+1) = \widehat{A}\widehat{x}(k) + \widehat{B}\widehat{u}(k) + \widehat{D}(k) \\ \widehat{y}(k) = \widehat{C}\widehat{x}(k) \end{cases} \tag{2.6}$$

where $\widehat{A} = e^{\overline{A}T}$, $\widehat{B} = \int_0^T e^{\overline{A}t}\overline{B}dt$, $\widehat{C} = \overline{C}$, and $\widehat{D}(k) = \int_0^T e^{\overline{A}t}\overline{D}d((k+1)T - t)dt$.

2.2.2 Estimation of Disturbance

In this part, we adopt the one-step delayed estimation strategy to estimate the disturbance $\widehat{D}(k)$ in Eq. (2.6), as follows:

$$\begin{aligned} \hat{\widehat{D}}(k) &= \widehat{D}(k-1) \\ &= \widehat{x}(k) - \widehat{A}\widehat{x}(k-1) - \widehat{B}\widehat{u}(k-1). \end{aligned} \tag{2.7}$$

Therefore, according to Eq. (2.6), we can deduce that:

$$\begin{cases} \widehat{x}(k+1) = \widehat{A}\widehat{x}(k) + \widehat{B}\widehat{u}(k) + \hat{\widehat{D}}(k) - \tilde{\widehat{D}}(k) \\ \widehat{y}(k) = \widehat{C}\widehat{x}(k) \end{cases} \tag{2.8}$$

where $\tilde{\widehat{D}}(k) = \hat{\widehat{D}}(k) - \widehat{D}(k)$ stands for the estimation error of disturbance.

Additionally, the error $\tilde{\widehat{D}}(k)$ can be modeled as:

$$\begin{aligned} \tilde{\widehat{D}}(k) &= \widehat{D}(k-1) - \widehat{D}(k) \\ &= \widehat{A}[\widehat{x}(k) - \widehat{x}(k-1)] + \widehat{B}[\widehat{u}(k) - \widehat{u}(k-1)] - [\widehat{x}(k+1) - \widehat{x}(k)] \\ &= \widehat{A}\Delta\widehat{x}(k) + \widehat{B}\Delta\widehat{u}(k) - \Delta\widehat{x}(k+1) \end{aligned} \tag{2.9}$$

where $\Delta\widehat{x}(k) = \widehat{x}(k) - \widehat{x}(k-1)$ and $\Delta\widehat{u}(k) = \widehat{u}(k) - \widehat{u}(k-1)$.

Assumption 2.1 It is assumed that the error of the disturbance estimation $\tilde{D}(k)$ is bounded, i.e.:

$$\left\| \tilde{\widehat{D}}(k) \right\| \leq \widehat{D}_{bou} \tag{2.10}$$

where $\|\cdot\|$ denotes the 2-norm of vector or matrix and $\widehat{D}_{bou} > 0$.

Remark 2 The control input $\tilde{\theta}$, acceleration $\tilde{a}$, and velocity $\tilde{v}$ are all bounded in an intelligent vehicle system, i.e., the state $\widehat{x}(k)$ and control input $\widehat{u}(k)$ are bounded in the plant (2.6). Based on Eq. (2.9), we know that the error $\widehat{\tilde{D}}(k)$ is also bounded. Therefore, Assumption 2.1 is reasonable.

2.3 Design of High-Order Adaptive Discrete-Time Sliding-Mode Control Method

2.3.1 Design of Sliding-Mode Function

In this part, we denote the tracking error as follows:

$$\widehat{e}(k) = \widehat{y}(k) - \widehat{y}_{ref}(k) \tag{2.11}$$

where $\widehat{y}_{ref}(k)$ denotes the intelligent vehicle's desired velocity.

Based on Eq. (2.11), we can construct the following sliding-mode function

$$\widehat{S}(k) = \widehat{S}(k-1) + \widehat{k}_1\widehat{e}(k) + \widehat{k}_2 sig^{\widehat{p}_1/\widehat{p}_2}(\Delta\widehat{e}(k-1)) \tag{2.12}$$

where $\widehat{k}_1 > 0$, $\widehat{k}_2 > 0$, $sig^{\widehat{p}_1/\widehat{p}_2}(\Delta\widehat{e}(k-1)) = \left\|\Delta\widehat{e}(k-1)\right\|^{\widehat{p}_1/\widehat{p}_2} sat(\Delta\widehat{e}(k-1))$, $\Delta\widehat{e}(k-1) = \widehat{e}(k) - \widehat{e}(k-1)$, $sat(\cdot)$ is the saturation function, $\widehat{p}_1$ and $\widehat{p}_2$ (satisfying the relation $0 < \widehat{p}_1/\widehat{p}_2 < 1$) are positive odd numbers.

Remark 3 In Eq. (2.12), we use the nonlinear term $sig^{\widehat{p}_1/\widehat{p}_2}(\Delta\widehat{e}(k-1))$ to improve the convergence speed, resulting in a steeper slope of the sliding surface in the neighborhood of the origin of the phase plane. Additionally, in Eq. (2.12), the saturation function is used to replace the traditional sign function to alleviate the chattering problem.

Furthermore, we propose the following high-order sliding-mode function:

$$\widehat{\delta}(k) = \widehat{S}(k) + \sum_{i=1}^{k-1} \widehat{w}_i[\widehat{S}(i) - (1-\widehat{q}T)\widehat{S}(i-1) + \widehat{\varepsilon}T\mathrm{sgn}(\widehat{S}(i-1))] \tag{2.13}$$

where $\widehat{w}_i$ denotes the adaptive factor with $\widehat{w}_i = \left\|\hat{\widehat{D}}(i)\right\| / \sum_{i=1}^{k-1} \left\|\hat{\widehat{D}}(i)\right\|$, $\widehat{q}$ is a converging parameter, and $\widehat{\varepsilon}$ stands for the switching parameter; both satisfy the relationship: $0 < 1 - \widehat{q}T < 1$, and $\text{sgn}(\cdot)$ presents the sign function.

Remark 4 Usually, the reaching law term $\widehat{S}(i-1) - \widehat{q}T\widehat{S}(i-1) - \widehat{\varepsilon}T\text{sgn}(\widehat{S}(i-1))$ is used as the equivalent value of $\widehat{S}(i)$ in DSMC [11]. Unlike in traditional methods, we use the accumulated weight deviations between the term $\widehat{S}(i-1) - \widehat{q}T\widehat{S}(i-1)$ $-\widehat{\varepsilon}T\text{sgn}(\widehat{S}(i-1))$ and the sliding mode function $\widehat{S}(i)$ as feedback signals to establish the high-order sliding mode function. It is worth noting that the overlay items consist of historical data. Therefore, it can be directly implemented.

Remark 5 In Eq. (2.13), we employ the estimated value of disturbance as feedback information to construct the high-order sliding mode function to suppress the effect of the disturbance. Additionally, we notice that the adaptive factor $\widehat{w}_i$ is determined by $\left\|\hat{\widehat{D}}(i)\right\|$. When the proportion of $\left\|\hat{\widehat{D}}(i)\right\|$ increases, the weight $\widehat{w}_i$ increases, meanwhile, the effect of ith term deviation becomes more prominent and vice versa. Obviously, $\sum_{i=1}^{k-1} \widehat{w}_i = 1$ and $0 \le \widehat{w}_i \le 1$.

Furthermore, let $\overline{\widehat{w}}_{k-1} = \overbrace{[\widehat{w}_1, \widehat{w}_2, \cdots, \widehat{w}_{k-1}]}^{k-1}$, $\overline{\widehat{S}}(k-1) = \overbrace{[\Delta\widehat{S}_1, \Delta\widehat{S}_2, \cdots, \Delta\widehat{S}_{k-1}]^T}^{k-1}$ and $\Delta\widehat{S}_i = \widehat{S}(i) - (1-\widehat{q}T) \times \widehat{S}(i-1) + \widehat{\varepsilon}T\text{sgn}(\widehat{S}(i-1))$. Then Eq. (2.13) can be translated into

$$\widehat{\delta}(k) = \widehat{S}(k) + \overline{\widehat{w}}_{k-1}\overline{\widehat{S}}(k-1) \tag{2.14}$$

2.3.2 Design of High-Order Sliding-Mode Controller

In this part, the equivalent control method is employed for designing the high-order adaptive sliding mode controller. Let $\widehat{\delta}(k+1) = 0$, it yields:

$$\widehat{\delta}(k+1) = \widehat{S}(k+1) + \overline{\widehat{w}}_k\overline{\widehat{S}}(k) = 0 \tag{2.15}$$

Therefore, it can be derived that

$$\widehat{S}(k+1) = -\overline{\widehat{w}}_k\overline{\widehat{S}}(k) \tag{2.16}$$

Based on Eq. (2.12), it is deduced that

$$\overline{\widehat{w}}_k \overline{\widehat{S}}(k) = -\widehat{S}(k) - \widehat{k}_1[\widehat{y}(k+1) - \widehat{y}_{ref}(k+1)] - \widehat{k}_2 sig^{\widehat{p}_1/\widehat{p}_2}(\Delta\widehat{e}(k)) \tag{2.17}$$

Furthermore, by substituting Eq. (2.8) into Eq. (2.17), we obtain:

$$\begin{aligned}\overline{\widehat{w}}_k \overline{\widehat{S}}(k) &= -\widehat{S}(k) - \widehat{k}_1[\widehat{C}(\widehat{A}\widehat{x}(k) + \widehat{B}\widehat{u}(k) + \hat{\widehat{D}}(k) \\ &\quad - \tilde{\widehat{D}}(k)) - \widehat{y}_{ref}(k+1)] - \widehat{k}_2 sig^{\widehat{p}_1/\widehat{p}_2}(\Delta\widehat{e}(k))\end{aligned} \tag{2.18}$$

When the estimation error of disturbance satisfies $\tilde{\widehat{D}}(k) = 0$, by solving (2.18), we obtain the following equivalent control $\widehat{u}_{eq}(k)$:

$$\begin{aligned}\widehat{u}_{eq}(k) &= -(\widehat{C}\widehat{B})^{-1}[\widehat{k}_1^{-1}\overline{\widehat{w}}_k \overline{\widehat{S}}(k) + \widehat{k}_1^{-1}\widehat{S}(k) \\ &\quad + \widehat{k}_2/\widehat{k}_1 sig^{\widehat{p}_1/\widehat{p}_2}(\Delta\widehat{e}(k)) + \widehat{C}\widehat{A}\widehat{x}(k) + \widehat{C}\hat{\widehat{D}}(k) - \widehat{y}_{ref}(k+1)]\end{aligned} \tag{2.19}$$

Furthermore, it can be deduced that

$$\begin{aligned}\widehat{u}_{eq}(k) = -(\widehat{C}\widehat{B})^{-1}\{&\widehat{k}_1^{-1}\sum_{i=1}^{k}\widehat{w}_i[\widehat{S}(i) - (1-\widehat{q}T)\widehat{S}(i-1) \\ &+ \widehat{\varepsilon}T\,\mathrm{sgn}(\widehat{S}(i-1))] + \widehat{k}_1^{-1}\widehat{S}(k) + \widehat{k}_2/\widehat{k}_1 sig^{\widehat{p}_1/\widehat{p}_2}(\Delta\widehat{e}(k)) \\ &+ \widehat{C}\widehat{A}\widehat{x}(k) + \widehat{C}\hat{\widehat{D}}(k) - \widehat{y}_{ref}(k+1)\}.\end{aligned} \tag{2.20}$$

Additionally, the following discontinuous term $\widehat{u}_{dis}(k)$ is added to guarantee the robustness [26]:

$$\widehat{u}_{dis}(k) = \widehat{u}_{dis}(k-1) - (\widehat{C}\widehat{B})^{-1}\widehat{\kappa} sign(\widehat{\sigma}(k)) \tag{2.21}$$

where $\widehat{\kappa} > 0$ stands for the gain of the switching control, and $sign(\cdot)$ is the sign function.

Hence, the total control input consists of the equivalent control $\widehat{u}_{eq}(k)$ and the switching control $\widehat{u}_{dis}(k)$:

$$\widehat{u}(k) = \widehat{u}_{eq}(k) + \widehat{u}_{dis}(k) \tag{2.22}$$

2.3.3 Analysis of Stability

Definition 2.1 For discrete-time systems, then the sliding surface is reachable if the following condition is satisfied [27]:

$$\widehat{\delta}(k+1)^2 < \widehat{\delta}(k)^2 \Rightarrow \left|\widehat{\delta}(k+1)\right| < \left|\widehat{\delta}(k)\right| \tag{2.23}$$

The following assumption is made to determine the reachability:

Assumption 2.2 The following inequality is held [28]:

$$\left|\widehat{k}_1\widehat{C}[\tilde{\widehat{D}}(k-1) - \tilde{\widehat{D}}(k)]\right| < \widehat{\hbar} < \widehat{k}_1\widehat{\kappa} \tag{2.24}$$

where $\widehat{\hbar} > 0$.

Theorem 2.1 *For the intelligent vehicle system defined in Eq.* (2.6) *with the Assumptions* 2.1*and* 2.2, *the sliding motion is reachable if the controller in Eq.* (2.22) *is used. Moreover, for any initial sliding surface* $\delta(0)$, *it can converge to the following quasi sliding mode domain in finite time:*

$$\widehat{\Xi} = \{\widehat{\delta}(k) : |\widehat{\delta}(k)| < \widehat{\hbar} + \widehat{k}_1\widehat{\kappa}\} \tag{2.25}$$

Proof The proof process can be divided into the following three steps.

(1) The sliding mode reachability

According to Eq. (2.13), the one-step prediction of high-order sliding-mode function is defined as:

$$\begin{aligned}\widehat{\delta}(k+1) &= \widehat{S}(k+1) + \sum_{i=1}^{k}\widehat{w}_i[\widehat{S}(i) - (1-\widehat{q}T)\widehat{S}(i-1) + \widehat{\varepsilon}T\operatorname{sgn}(\widehat{S}(i-1))]\\ &= \widehat{S}(k) + \widehat{k}_1\widehat{e}(k+1) + \widehat{k}_2 sig^{\widehat{p}_1/\widehat{p}_2}(\Delta\widehat{e}(k)) + \sum_{i=1}^{k}\widehat{w}_i[\widehat{S}(i)\\ &\quad - (1-\widehat{q}T)\widehat{S}(i-1) + \widehat{\varepsilon}T\operatorname{sgn}(\widehat{S}(i-1))]\end{aligned} \tag{2.26}$$

According to Eq. (2.11), (2.8), (2.19), and (2.21), the following is obtained:

$$\begin{aligned}\widehat{e}(k+1) &= \widehat{y}(k+1) - \widehat{y}_{ref}(k+1)\\ &= \widehat{C}\widehat{x}(k+1) - \widehat{y}_{ref}(k+1)\end{aligned}$$

$$
\begin{aligned}
&= \widehat{C}[\widehat{A}\widehat{x}(k) + \widehat{B}\widehat{u}(k) + \hat{\widehat{D}}(k) - \tilde{\widehat{D}}(k)] - \widehat{y}_{ref}(k+1) \\
&= \widehat{C}\widehat{A}\widehat{x}(k) - [\widehat{k}_1^{-1}\overline{\widehat{w}}_k\overline{\widehat{S}}(k) + \widehat{k}_1^{-1}\widehat{S}(k) \\
&\quad + \widehat{k}_2/\widehat{k}_1 sig^{\widehat{p}_1/\widehat{p}_2}(\Delta\widehat{e}(k)) + \widehat{C}\widehat{A}\widehat{x}(k) + \widehat{C}\hat{\widehat{D}}(k) \\
&\quad - \widehat{y}_{ref}(k+1)] + \widehat{C}\widehat{B}u_{dis}(k) + \widehat{C}\hat{\widehat{D}}(k) - \widehat{C}\tilde{\widehat{D}}(k) - \widehat{y}_{ref}(k+1) \\
&= -\widehat{k}_1^{-1}\overline{\widehat{w}}_k\overline{\widehat{S}}(k) - \widehat{k}_1^{-1}\widehat{S}(k) - \widehat{k}_2/\widehat{k}_1 sig^{\widehat{p}_1/\widehat{p}_2}(\Delta\widehat{e}(k)) \\
&\quad + \widehat{C}\widehat{B}u_{dis}(k) - \widehat{C}\tilde{\widehat{D}}(k)
\end{aligned} \tag{2.27}
$$

By substituting Eq. (2.27) into Eq. (2.26), we derive:

$$
\widehat{\delta}(k+1) = \widehat{k}_1\widehat{C}\widehat{B}u_{dis}(k) - \widehat{k}_1\widehat{C}\tilde{\widehat{D}}(k) \tag{2.28}
$$

Therefore, the following is obtained from Eq. (2.28):

$$
\widehat{\delta}(k) = \widehat{k}_1\widehat{C}\widehat{B}\widehat{u}_{dis}(k-1) - \widehat{k}_1\widehat{C}\tilde{\widehat{D}}(k-1) \tag{2.29}
$$

According to Assumption 2.1, we can conclude that $\widehat{\delta}(k)$ is bounded. Hence, the difference between Eqs. (2.28) and (2.29) yields:

$$
\begin{aligned}
\widehat{\delta}(k+1) - \widehat{\delta}(k) &= \widehat{k}_1\widehat{C}\widehat{B}[\widehat{u}_{dis}(k) - \widehat{u}_{dis}(k-1)] - \widehat{k}_1\widehat{C}[\tilde{\widehat{D}}(k) - \tilde{\widehat{D}}(k-1)] \\
&= -\widehat{k}_1\widehat{\kappa}sign(\widehat{\sigma}(k)) + \widehat{k}_1\widehat{C}[\tilde{\widehat{D}}(k-1) - \tilde{\widehat{D}}(k)]
\end{aligned} \tag{2.30}
$$

Considering Assumption 2.2, it is deduced that:

$$
-\widehat{k}_1\widehat{\kappa}sign(\widehat{\sigma}(k)) - \widehat{\hbar} < \widehat{\delta}(k+1) - \widehat{\delta}(k) < -\widehat{k}_1\widehat{\kappa}sign(\widehat{\sigma}(k)) + \widehat{\hbar} \tag{2.31}
$$

(i) For any $k \in z^+$, if the inequality $\widehat{\delta}(k) > 0$ holds, i.e., the sequence $\{\widehat{\delta}(k), k \in z^+\}$ is positive, we can conclude from Eq. (2.31) that:

$$
\widehat{\delta}(k+1) - \widehat{\delta}(k) < -\widehat{k}_1\widehat{\kappa} + \widehat{\hbar} < 0 \tag{2.32}
$$

Equation (2.32) shows that $\widehat{\delta}(k+1) < \widehat{\delta}(k) \Rightarrow |\widehat{\delta}(k+1)| < |\widehat{\delta}(k)|$.

(ii) For any $k \in z^+$, if the inequality $\widehat{\delta}(k) < 0$ holds, it can be obtained from Eq. (2.31) that:

$$0 < \widehat{k}_1\widehat{\kappa} - \widehat{\hbar} < \widehat{\delta}(k+1) - \widehat{\delta}(k) \tag{2.33}$$

Equation (2.33) shows that $\widehat{\delta}(k+1) > \widehat{\delta}(k)$ because of $\widehat{\delta}(k) < 0$; thus, $\left|\widehat{\delta}(k+1)\right| < \left|\widehat{\delta}(k)\right|$. Synthesizing (i) and (ii), it can be concluded that:

$$\left|\widehat{\delta}(k+1)\right| < \left|\widehat{\delta}(k)\right| \tag{2.34}$$

Based on Definition 2.1, the reachability of the sliding surface can be guaranteed.

(2) Convergence in finite time

Next, we will prove the finite-time convergence, i.e., for any initial value $\widehat{\delta}(0)$, it will converge to the sliding domain $\widehat{\Xi}$ in finite time.

(i) When the initial value $\widehat{\delta}(0) > \widehat{\hbar} + \widehat{k}_1\widehat{\kappa}$, the distance between the quasi-sliding mode domain and the initial value can be computed by $\widehat{\delta}(0) - \widehat{\hbar} + \widehat{k}_1\widehat{\kappa}$. We define the minimum move step as $\widehat{s}_{\min}$. It is worth noting that $\widehat{\delta}(k) - \widehat{\delta}(k+1) > \widehat{k}_1\widehat{\kappa} - \widehat{\hbar} > \widehat{s}_{\min}$; therefore, it can be derived that:

$$\widehat{k}^* = (\widehat{\delta}(0) - \widehat{\hbar} + \widehat{k}_1\widehat{\kappa})/\widehat{s}_{\min} < (\widehat{\delta}(0) - \widehat{\hbar} + \widehat{k}_1\widehat{\kappa})/(\widehat{k}_1\widehat{\kappa} - \widehat{\hbar}) \tag{2.35}$$

It is noted that $\widehat{k}^*$ is a finite number.

(ii) When $\widehat{\delta}(0) < -\widehat{\hbar} - \widehat{k}_1\widehat{\kappa}$, a similar result can be presented.

Synthesizing (i) and (ii), we can conclude that any initial value $\widehat{\delta}(0)$ can converge to the sliding domain $\widehat{\Xi}$ in finite time.

(3) Remaining in the sliding domain after time $\widehat{k}^*$.

Next, we prove that $\widehat{\delta}(k)(k \geq \widehat{k}*)$ remains in the sliding domain, i.e., $\widehat{\delta}(k) \in \widehat{\Xi}(k \geq \widehat{k}*)$ holds. Here, we employ the mathematical induction method, and the value $\widehat{\delta}(\widehat{k}*)$ is considered as the first certificate induction term. The proof process is as follows:

(1) If $n = \widehat{k}*$, $\widehat{\delta}(\widehat{k}*) \in \widehat{\Xi}$ is held.

(2) When $n = k(k > \widehat{k}*)$, then $\left|\widehat{\delta}(n)\right| < \widehat{\hbar} + \widehat{k}_1\widehat{\kappa}$, i.e., $\widehat{\delta}(n) \in \widehat{\Xi}$ is held. In the following, we prove that

$$\left|\widehat{\delta}(n)\right| < \widehat{\hbar} + \widehat{k}_1\widehat{\kappa}(n = k+1) \tag{2.36}$$

(i) When the value of $\widehat{\delta}(k)$ is positive, i.e., $0 < \widehat{\delta}(k) < \widehat{\hbar} + \widehat{k}_1\widehat{\kappa}$. Based on Eq. (2.30), it is deduced that:

$$\widehat{\delta}(k+1) = \widehat{\delta}(k) - \widehat{k}_1\widehat{\kappa} + \widehat{k}_1\widehat{C}[\widetilde{\widehat{D}}(k-1) - \widetilde{\widehat{D}}(k)] \tag{2.37}$$

According to Assumption 2.2, we can obtain the following relation:

$$\widehat{\delta}(k) - \widehat{k}_1\widehat{\kappa} - \widehat{\hbar} < \widehat{\delta}(k+1) < \widehat{\delta}(k) - \widehat{k}_1\widehat{\kappa} + \widehat{\hbar} \tag{2.38}$$

Moreover, the following is obtained:

$$-\widehat{k}_1\widehat{\kappa} - \widehat{\hbar} < \widehat{\delta}(k+1) < 2\widehat{\hbar} < \widehat{k}_1\widehat{\kappa} + \widehat{\hbar} \tag{2.39}$$

Inequality (2.39) shows that $\widehat{\delta}(k+1) \in \widehat{\Xi}$.

(ii) When $-\widehat{\hbar} - \widehat{k}_1\widehat{\kappa} < \widehat{\delta}(k) < 0$, i.e., the value of $\widehat{\delta}(k)$ is negative, a similar result can be obtained.

Synthesizing (i) and (ii), we determine that the inequality in (2.36) is held. Based on mathematical induction, we can conclude that $\widehat{\delta}(k)(k \geq \widehat{k}*)$ remains in the domain $\widehat{\Xi}$, i.e., $\widehat{\delta}(k) \in \widehat{\Xi}$ for $\forall k \geq \widehat{k}^*$. This completes the proof.

2.4 Control of Networked Intelligent Vehicle System

In this part, the control of the networked intelligent vehicle system is analyzed, and the proposed controller design is implemented. Here, we assume that the feedback channel experiences network-induced disturbance, whereas the feedforward channel does not.

The sampling output can be rewritten in the following form due to the existence of network-induced disturbance in the feedback channel:

$$\overline{\widehat{y}}(k) = \widehat{y}(k) + \widehat{y}_d(k) \tag{2.40}$$

where $\widehat{y}(k)$ is the sampling output, $\overline{\widehat{y}}(k)$ is the actual received signal of the controller, and $\widehat{y}_d(k)$ is the network-induced disturbance.

Remark 6 Generally, the disturbance $\widehat{y}_d(k)$ often originates from sensor faults, measurement noise, and hacker attacks (for example, false data injection attacks).

Assumption 2.3 The disturbance $\widehat{y}_d(k)$ is bounded.

2.4.1 Compensation of Network-Induced Disturbance

As shown in Fig. 2.1, an observer on the side of the control end is designed for dealing with the network-induced disturbance term $\widehat{y}_d(k)$. Since the existence of network-induced disturbance is in the feedback channel, the description of the discrete-time plant in Eq. (2.6) can be rewritten as

$$\begin{cases} \widehat{x}(k+1) = \widehat{A}\widehat{x}(k) + \widehat{B}\widehat{u}(k) + \widehat{D}(k) \\ \widehat{\overline{y}}(k) = \widehat{y}(k) + \widehat{y}_d(k). \end{cases} \tag{2.41}$$

Let $\widehat{\overline{x}}(k) = [\widehat{x}^T(k)\,\widehat{y}_d(k)]^T$, $\widehat{\overline{E}} = [I_2 0_{2\times1}]$, $\widehat{\overline{A}} = [\widehat{A}\, 0_{2\times1}]$, $\widehat{\overline{C}} = [\widehat{C}1]$. Then, we can obtain a new augmented system as follows:

$$\begin{cases} \widehat{\overline{E}}\widehat{\overline{x}}(k+1) = \widehat{\overline{A}}\widehat{\overline{x}}(k) + \widehat{B}\widehat{u}(k) + \widehat{D}(k) \\ \widehat{\overline{y}}(k) = \widehat{\overline{C}}\widehat{\overline{x}}(k) \end{cases} \tag{2.42}$$

Additionally, let $\widehat{\Psi} = \begin{bmatrix} I_2 & -\widehat{C}^T \end{bmatrix}^T$ and $\widehat{\Theta} = \begin{bmatrix} 0_{2\times1}^T & I_1 \end{bmatrix}^T$. In this part, in the augmented system (2.42), we design the following observer:

$$\begin{cases} \widehat{\vartheta}(k+1) = (\widehat{\Psi}\widehat{\overline{A}} - \widehat{L}\widehat{\overline{C}})\widehat{\vartheta}(k) + [\widehat{L} + (\widehat{\Psi}\widehat{\overline{A}} - \widehat{L}\widehat{\overline{C}})\widehat{\Theta}]\widehat{\overline{y}}(k) + \widehat{\Psi}\widehat{B}\widehat{u}(k) + \widehat{u}_d(k) \\ \hat{\widehat{\overline{x}}}(k) = \widehat{\vartheta}(k) + \widehat{\Theta}\widehat{\overline{y}}(k) \end{cases} \tag{2.43}$$

where $\hat{\widehat{\overline{x}}}(k)$ is the estimated state of $\widehat{\overline{x}}(k)$, $\widehat{\vartheta}(k)$ is the intermediate variables, $\widehat{L}$ is the observer gain to be determined, and $\widehat{u}_d(k)$ is the discontinuous input term.

Remark 7 In the observer (2.43), the input term $\widehat{u}_d(k)$ is added to suppress the external disturbance $\widehat{D}(k)$ in Eq. (2.42) and obtain sufficient accuracy.

We provide the following theorem before solving the gain of the observer.

Theorem 2.2 *For $\widehat{D}(k) = 0$, if we can select a matrix $\widehat{L}$ so that $\widehat{\Psi}\widehat{\overline{A}} - \widehat{L}\widehat{\overline{C}}$ is stable, the estimation error $\widehat{\overline{e}}_x(k) = \widehat{\overline{x}}(k) - \hat{\widehat{\overline{x}}}(k)$ can asymptotically approach to 0.*

Proof For the first equation in (2.42), we obtain the following before and after multiplying $\widehat{\Psi}$:

$$\widehat{\Psi}\widehat{\overline{E}}\widehat{\overline{x}}(k+1) = \widehat{\Psi}\widehat{\overline{A}}\widehat{\overline{x}}(k) + \widehat{\Psi}\widehat{B}\widehat{u}(k) + \widehat{\Psi}\widehat{D}(k) \tag{2.44}$$

We add $\widehat{\Theta}\widehat{\overline{y}}(k+1)$ to both sides of Eq. (2.44):

$$\overline{x}(k+1) = \widehat{\Psi}\widehat{\overline{A}}\overline{x}(k) + \widehat{\Psi}\widehat{B}\widehat{u}(k) + \widehat{\Theta}\widehat{\overline{y}}(k+1) + \widehat{\Psi}\widehat{D}(k) \tag{2.45}$$

Let $\widehat{u'}(k) = \widehat{\Psi}\widehat{B}\widehat{u}(k) + \widehat{\Theta}\widehat{\overline{y}}(k+1)$; then, the plant (2.42) can be rewritten as:

$$\begin{cases} \overline{x}(k+1) = \widehat{\Psi}\widehat{\overline{A}}\overline{x}(k) + \widehat{u'}(k) + \widehat{\Psi}\widehat{D}(k) \\ \widehat{\overline{y}}(k) = \widehat{\overline{C}}\overline{x}(k) \end{cases} \tag{2.46}$$

For system (2.46), we design the following Luenberger observer:

$$\begin{aligned} \hat{\overline{x}}(k+1) &= \widehat{\Psi}\widehat{\overline{A}}\hat{\overline{x}}(k) + \widehat{u'}(k) + \widehat{L}(\widehat{\overline{y}}(k) - \widehat{\overline{C}}\hat{\overline{x}}(k)) \\ &= \widehat{\Psi}\widehat{\overline{A}}\hat{\overline{x}}(k) + \widehat{\Psi}\widehat{B}\widehat{u}(k) + \widehat{\Theta}\widehat{\overline{y}}(k+1) + \widehat{L}(\widehat{\overline{y}}(k) - \widehat{\overline{C}}\hat{\overline{x}}(k)) \end{aligned} \tag{2.47}$$

where $\widehat{L}$ denotes the gain of the observer, which will be described later.

Let $\widehat{\vartheta}(k) = \hat{\overline{x}}(k) - \hat{\Theta}\hat{\overline{y}}(k)$; then Eq. (2.47) can be rewritten as:

$$\widehat{\vartheta}(k+1) = (\widehat{\Psi}\widehat{\overline{A}} - \widehat{L}\widehat{\overline{C}})\widehat{\vartheta}(k) + [\widehat{L} + (\widehat{\Psi}\widehat{\overline{A}} - \widehat{L}\widehat{\overline{C}})\widehat{\Theta}]\widehat{\overline{y}}(k) + \widehat{\Psi}\widehat{B}\widehat{u}(k) \tag{2.48}$$

We find that the parameters of the observer in Eq. (2.43) are the as same as in Eq. (2.48).

If $\widehat{D}(k) = 0$, we subtract Eq. (2.45) from Eq. (2.47) to obtain:

$$\widehat{\overline{e}}_x(k+1) = (\widehat{\Psi}\widehat{\overline{A}} - \widehat{L}\widehat{\overline{C}})\widehat{\overline{e}}_x(k) \tag{2.49}$$

Equation (2.49) shows that the error $\widehat{\overline{e}}_x(k)$ asymptotically converges to 0 if $\widehat{\Psi}\widehat{\overline{A}} - \widehat{L}\widehat{\overline{C}}$ is stable. This completes the proof.

Remark 8 It is worth noting that when $\widehat{D}(k) = 0$, the gain of the observer $\widehat{L}$ can be selected to ensure that $\widehat{\Psi}\widehat{\overline{A}} - \widehat{L}\widehat{\overline{C}}$ is stable. In the following, the input term $\widehat{u}_d(k)$ is used to reject the disturbance term $\widehat{D}(k)$.

In this part, the DSMC is employed to construct the input term $\widehat{u}_d(k)$. First, the sliding mode function is defined as follows:

$$\widehat{S}_x(k) = \widehat{\alpha}\widehat{\overline{e}}_x(k) + \widehat{\beta}(\Delta\widehat{\overline{e}}_x(k))^{1/2} \tag{2.50}$$

where $\Delta\widehat{e}_x(k) = \widehat{e}_x(k) - \widehat{e}_x(k-1)$, and $\widehat{\alpha}$ and $\widehat{\beta}$ are two tuning parameters.

Then, the input $\widehat{u}_d(k)$ is constructed as

$$\widehat{u}_d(k) = -\widehat{\Im} \cdot sign(\widehat{S}_x(k)) \tag{2.51}$$

where $\widehat{\Im}$ denotes the parameter to be estimated.

It should be noted that $\hat{\widehat{x}}(k) = [\hat{\widehat{x}}(k)\ \hat{\widehat{y}}_d(k)]^T$, and $\hat{\widehat{y}}(k) = \widehat{C}\hat{\widehat{x}}(k)$. The observer delivers the estimated output to the control terminal. In the following, we redesign the controller according to the observer (2.43) to suppress the network-induced disturbance $\widehat{y}_d(k)$. Based on Eq. (2.11), the error of the tracking output is redefined as:

$$\widehat{e}_y(k) = \hat{\widehat{y}}(k) - \widehat{y}_{ref}(k) \tag{2.52}$$

Hence, we can design the following sliding mode function:

$$\widehat{S}_y(k) = \widehat{S}_y(k-1) + \widehat{k}_1\widehat{e}_y(k) + \widehat{k}_2 sig^{\widehat{p}_1/\widehat{p}_2}(\Delta\widehat{e}_y(k-1)) \tag{2.53}$$

Furthermore, the equivalent control input can be obtained as:

$$\begin{aligned}\widehat{u}^{y}_{eq}(k) = &-(\widehat{C}\widehat{B})^{-1}\{\widehat{k}_1^{-1}\sum_{i=1}^{k}\widehat{\omega}_i[\widehat{S}_y(i) - (1-\widehat{q}T)\widehat{S}_y(i-1)\\ &+\widehat{\varepsilon}T\mathrm{sgn}(\widehat{S}_y(i-1))] + \widehat{k}_1^{-1}\widehat{S}_y(k)\\ &+\widehat{k}_2/\widehat{k}_1 sig^{\widehat{p}_1/\widehat{p}_2}(\Delta\widehat{e}_y(k)) + \widehat{C}\widehat{A}\hat{\widehat{x}}(k)\\ &+\widehat{C}\hat{\widehat{D}}(k) - \widehat{y}_{ref}(k+1)\}.\end{aligned} \tag{2.54}$$

Then, the controller transmits the output $\widehat{u}(k) = \widehat{u}^{y}_{eq}(k) + \widehat{u}_{dis}(k)$ to the actuator side, and the actuator powers the vehicle.

2.4.2 *Tracking Error Analysis*

In this part, the error of the tracking output is discussed.

Assumption 2.4 The nonlinear function $sig^{\widehat{p}_1/\widehat{p}_2}(\Delta\widehat{e}_y(k))$ is bounded, i.e.,$\left|sig^{\widehat{p}_1/\widehat{p}_2}(\Delta\widehat{e}_y(k))\right| < \widehat{\Delta}_2$, where $\widehat{\Delta}_2 > 0$ is a constant.

Theorem 2.3 *We consider the plant (2.41) under control input $\widehat{u}(k)$. If Assumption 2.4holds, the error of the tracking output will converge to a domain.*

Proof Based on Eq. (2.52), the error of the tracking output is expressed as:

$$\begin{aligned}\widehat{e}_y(k+1) &= \hat{\widehat{y}}(k+1) - \widehat{y}_{ref}(k+1)\\ &= \widehat{C}\hat{\widehat{x}}(k+1) - \widehat{y}_{ref}(k+1)\\ &= \widehat{C}(\widehat{A}\hat{\widehat{x}}(k) + \widehat{B}\widehat{u}(k) + \widehat{D}(k)) - \widehat{y}_{ref}(k+1).\end{aligned}$$

Additionally, we obtain the following due to the control input in Eq. (2.54):

$$\begin{aligned}\widehat{e}_y(k+1) &= \widehat{C}\widehat{A}\hat{\widehat{x}}(k) + \widehat{C}\widehat{B}(\widehat{u}_{dis}(k) + \widehat{u}^{y}_{eq}(k)) + \widehat{C}\widehat{D}(k) - \widehat{y}_{ref}(k+1)\\ &= \widehat{C}\widehat{A}\hat{\widehat{x}}(k) + \widehat{C}\widehat{B}\widehat{u}_{dis}(k) + [-\widehat{k}_1^{-1}\sum_{i=1}^{k}\widehat{w}_i[\widehat{S}_y(i) - (1-\widehat{q}T)\widehat{S}_y(i-1)\\ &+ \widehat{\varepsilon}T\,\mathrm{sgn}(\widehat{S}_y(i-1))] - \widehat{k}_1^{-1}\widehat{S}_y(k) - \widehat{k}_2/\widehat{k}_1 sig^{\widehat{p}_1/\widehat{p}_2}(\Delta\widehat{e}_y(k))\\ &- \widehat{C}\widehat{A}\hat{\widehat{x}}(k) - \widehat{C}\hat{\widehat{D}}(k) + \widehat{y}_{ref}(k+1)] + \widehat{C}\widehat{D}(k) - \widehat{y}_{ref}(k+1)\\ &= \widehat{C}\widehat{B}u_{dis}(k) - \widehat{k}_1^{-1}\sum_{i=1}^{k}\widehat{w}_i[\widehat{S}_y(i) - (1-\widehat{q}T)\widehat{S}_y(i-1)\\ &+ \widehat{\varepsilon}T\,\mathrm{sgn}(\widehat{S}_y(i-1))] - \widehat{k}_1^{-1}\widehat{S}_y(k)\\ &- \widehat{k}_2/\widehat{k}_1 sig^{\widehat{p}_1/\widehat{p}_2}(\Delta\widehat{e}_y(k)) - \widehat{C}\tilde{\widehat{D}}(k).\end{aligned}$$

Based on Eq. (2.28), we deduce:

$$\begin{aligned}\widehat{e}_y(k+1) = -\widehat{k}_1^{-1}\sum_{i=1}^{k}\widehat{w}_i[\widehat{S}_y(i) - (1-\widehat{q}T)\widehat{S}_y(i-1)\\ + \widehat{\varepsilon}T\,\mathrm{sgn}(\widehat{S}_y(i-1))] - \widehat{k}_1^{-1}\widehat{S}_y(k)\\ - \widehat{k}_2/\widehat{k}_1 sig^{\widehat{p}_1/\widehat{p}_2}(\Delta\widehat{e}_y(k)) + \widehat{k}_1^{-1}\widehat{\delta}(k+1)\end{aligned}$$

Since $\widehat{\delta}_y(k) = \overline{\widehat{S}}_y(k) + \widehat{w}_{k-1}\overline{\widehat{S}}_y(k-1)$ we can deduce that:

$$\begin{aligned}\widehat{e}_y(k+1) = -\widehat{k}_1^{-1}\widehat{w}_k[\widehat{S}_y(k) - (1-\widehat{q}T)\widehat{S}_y(k-1) + \widehat{\varepsilon}T\,\mathrm{sgn}(\widehat{S}_y(k-1))]\\ + \widehat{k}_1^{-1}[\widehat{\delta}(k+1) - \widehat{\delta}_y(k)] - \widehat{k}_2/\widehat{k}_1 sig^{\widehat{p}_1/\widehat{p}_2}(\Delta\widehat{e}_y(k))\end{aligned} \quad (2.55)$$

If k is sufficiently large, the following condition is satisfied:

$$\widehat{S}_y(k) - (1 - \widehat{q}T)\widehat{S}_y(k-1) + \widehat{\varepsilon}T\mathrm{sgn}(\widehat{S}_y(k-1)) \to 0 \tag{2.56}$$

That is:

$$\left|\widehat{k}_1^{-1}\widehat{w}_k[\widehat{S}_y(k) - (1 - \widehat{q}T)\widehat{S}_y(k-1) + \widehat{\varepsilon}T\mathrm{sgn}(\widehat{S}_y(k-1))]\right| < \widehat{k}_1^{-1}\widehat{\Delta}_1 \tag{2.57}$$

where $\widehat{\Delta}_1$ is a very small number.

According to Theorem 2.1, we can conclude that:

$$\left|\widehat{\delta}_y(k)\right| < \widehat{\hbar} + \widehat{k}_1\widehat{\kappa} \tag{2.58}$$

Hence, it can be derived that:

$$\left|\widehat{k}_1^{-1}[\widehat{\delta}(k+1) - \widehat{\delta}_y(k)]\right| < 2\widehat{k}_1^{-1}(\widehat{\hbar} + \widehat{k}_1\widehat{\kappa}) \tag{2.59}$$

According to Assumption 2.4, we obtain the following:

$$\left|\widehat{k}_1^{-1}\widehat{k}_2 sig^{\widehat{p}_1/\widehat{p}_2}(\Delta\widehat{e}_y(k))\right| < \widehat{k}_1^{-1}\widehat{k}_2\widehat{\Delta}_2 \tag{2.60}$$

Equations (2.57), (2.59), and (2.60) let us conclude that $\widehat{e}_y(k+1)$ is bounded. Therefore, $\widehat{e}_y(k+1)$ can converge to a domain. This completes the proof. □

2.5 Simulation Example

In this part, a simulation is carried out to validate the effectiveness of the proposed approaches. As presented in Eq. (2.4), the engine time constant is $\varsigma = 0.25$ [29]. By substituting this value into Eq. (2.4), we can obtain the following state-space equation:

$$\overline{A} = \begin{pmatrix} 0 & 1 \\ 0 & -4 \end{pmatrix}, \overline{B} = \begin{pmatrix} 0 \\ 4 \end{pmatrix}, \text{ and } \overline{C} = \begin{pmatrix} 1 & 0 \end{pmatrix}.$$

We use the following two simulation cases to demonstrate the effectiveness of the proposed observer for compensating for communication constraints and of the controller for suppressing the external disturbance. (1) The network is intact, and only interference occurs. (2) Two types of interferences (network-induced disturbance and external disturbance) occur simultaneously.

2.5.1 Case 1: The Network is Intact

It is assumed that the network is intact, i.e., there is no network-induced interference, and we only consider external disturbances as follows:

$$D(k) = 0.5\pi[\cos(2kT) + \sin(3kT)] \tag{2.61}$$

where k denotes the discrete instant.

We select the sampling period $T = 0.04$ and use a discretization approach to obtain the following discrete-time state equation:

$$\widehat{A} = \begin{pmatrix} 1 & 0.0370 \\ 0 & 0.8521 \end{pmatrix}, \widehat{B} = \begin{pmatrix} 0.0030 \\ 0.1479 \end{pmatrix}, \text{ and } \widehat{C} = \begin{pmatrix} 1 & 0 \end{pmatrix}.$$

The parameters of the controller are $\widehat{q} = 5$, $\widehat{\varepsilon} = 0.01$, $\widehat{p}_1 = 2$, $\widehat{p}_2 = 3$,$\widehat{k}_1 = 5$, $\widehat{k}_2 = 0.1$, and $\widehat{\kappa} = 0.01$. The initial values of the control input and system state are $\widehat{u}_0 = 0.5$ and $\widehat{x}_0 = \begin{pmatrix} 0.1 & 0.2 \end{pmatrix}^T$, respectively. The simulation time is 10 s, and the desired velocity of the vehicle is $\widehat{y}_{ref} = 100\,\text{km/h}$. The method presented in [12] is used as a comparison to demonstrate the effectiveness of the proposed methods. The simulation results are as follows.

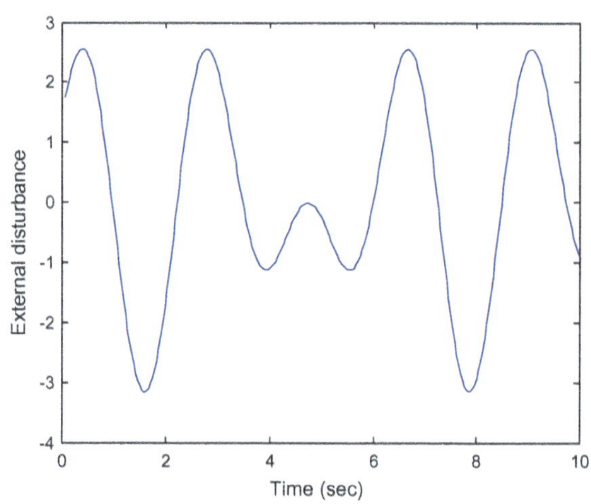

Fig. 2.2 The trajectory of external disturbance

Figure 2.2 presents the trajectory of the external disturbance, and Fig. 2.3 plots the trajectory of the tracking output for the vehicle velocity. The results indicate that the proposed approach can track the desired velocity for about 0.5 s; however, the control method presented in [12] requires about 1.5 s to achieve the same effect. Figure 2.4 shows the trajectory of the sliding mode surface; the results indicate that both methods can converge to zero in finite time. Figure 2.5 presents the trajectory of the control input. It can be seen that the proposed method has less chattering than the method proposed in [12].

Generally, when the vehicle is in motion, the velocity changes over time. Therefore, we assume that the desired velocity changes over time within 20 s as follows.

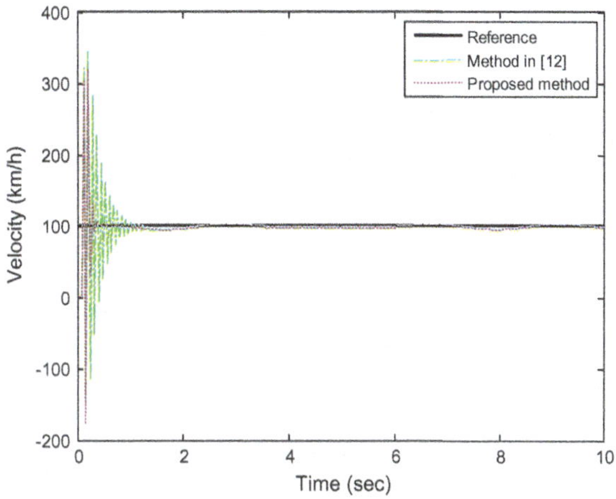

Fig. 2.3 The trajectory of tracking output for the velocity of vehicle

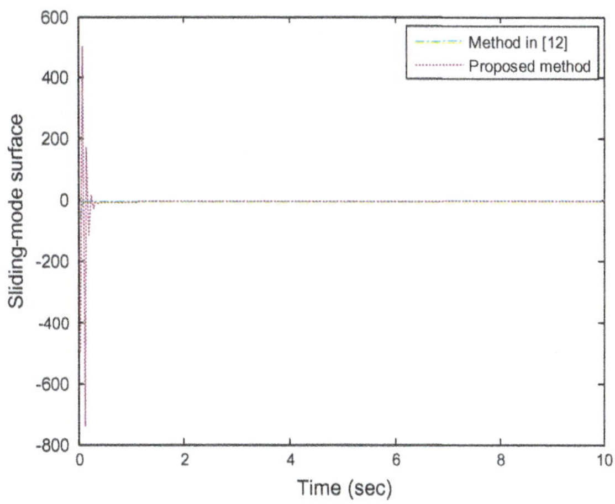

Fig. 2.4 The trajectory of the sliding-mode surface

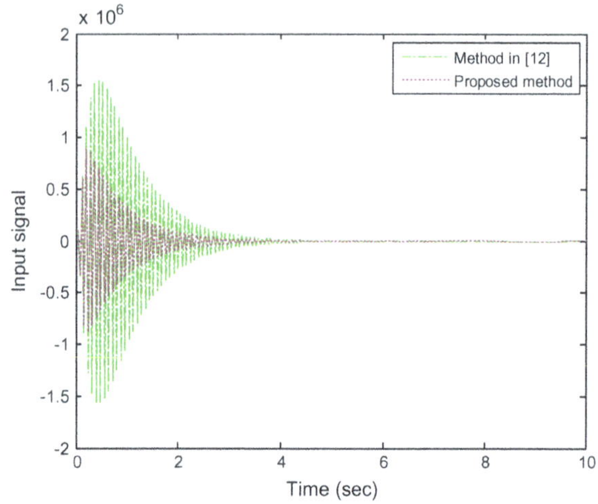

Fig. 2.5 The trajectory of control input

$$\widehat{y}_{ref} = \begin{cases} 30\,\text{km/h}, 0\,\text{s} < t \leq 4\,\text{s} \\ 60\,\text{km/h}, 4\,\text{s} < t \leq 8\,\text{s} \\ 90\,\text{km/h}, 8\,\text{s} < t \leq 12\,\text{s} \\ 100\,\text{km/h}, 12\,\text{s} < t \leq 16\,\text{s} \\ 120\,\text{km/h}, 16\,\text{s} < t \leq 20\,\text{s} \end{cases}.$$

We use the following external disturbances:

$$D(k) = 0.3\pi[\cos(4kT) + \sin(5kT)] \tag{2.62}$$

The other parameters are the same as above.

Figure 2.6 presents the trajectory of the external disturbance, and Fig. 2.7 plots the trajectory of the tracking output for the vehicle velocity. The results show that the proposed approach can track the desired velocity despite the time-varying speed. Figure 2.8 shows the trajectory of the sliding mode surface; the results indicate that both methods can converge to zero in finite time. Figure 2.9 presents the trajectory of the control input. It is observed that the proposed method has less jitter than the method proposed in [12].

2.5.2 *Case 2: Network-Induced Disturbance*

In this part, we discuss the case of network-induced disturbance. The controller parameters and the initial values are the same as above, and the observer parameters are $\widehat{\alpha} = 0.1$, $\widehat{\beta} = 0.2$ and $\widehat{\Im} = 0.5$. The desired velocity of the vehicle is

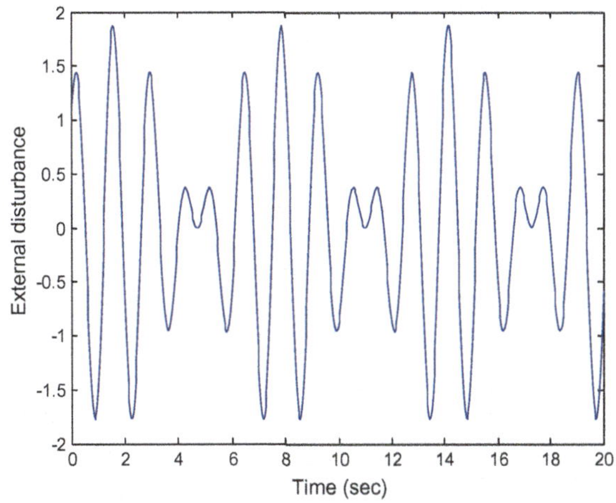

Fig. 2.6 The trajectory of external disturbance

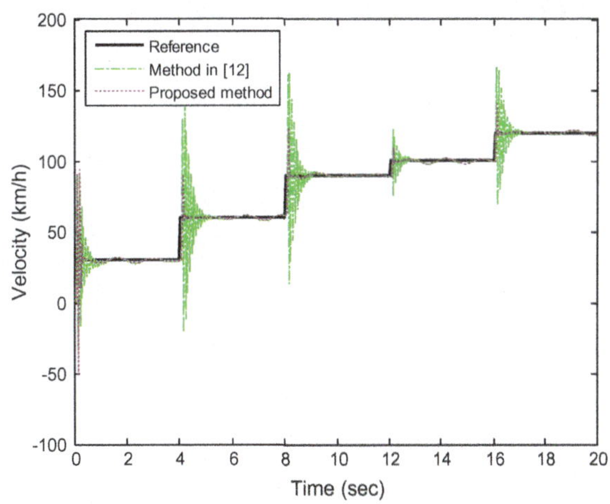

Fig. 2.7 The trajectory of tracking output for the velocity of the vehicle

$$\widehat{y}_{ref} = \begin{cases} 24t(\text{km/h}), & 0 < t \leq 5 \\ 120 - 8(t-5)(\text{km/h}), & 5 < t \leq 10 \\ 80, & 10 < t \leq 15 \\ 80 - 16(t-15)(\text{km/h}), & 15 < t \leq 20 \end{cases}.$$

The sampling period is $T = 0.05$, and the discrete-time state equation is obtained as follows:

$$\widehat{A} = \begin{pmatrix} 1 & 0.0453 \\ 0 & 0.8187 \end{pmatrix}, \widehat{B} = \begin{pmatrix} 0.0047 \\ 0.1813 \end{pmatrix} \text{and } \widehat{C} = \begin{pmatrix} 1 & 0 \end{pmatrix}.$$

We use the following external disturbance:

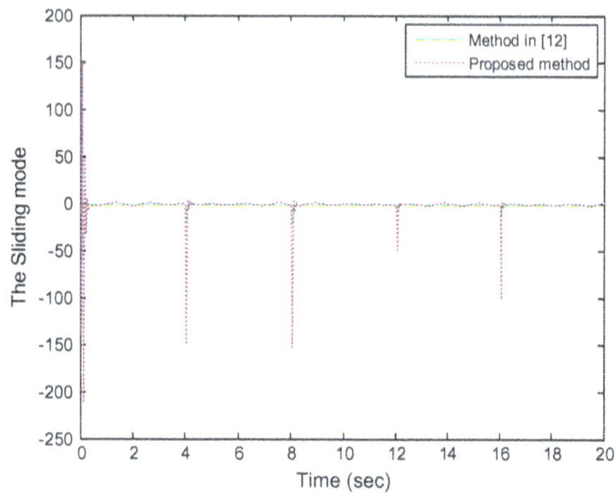

Fig. 2.8 The trajectory of the sliding-mode surface

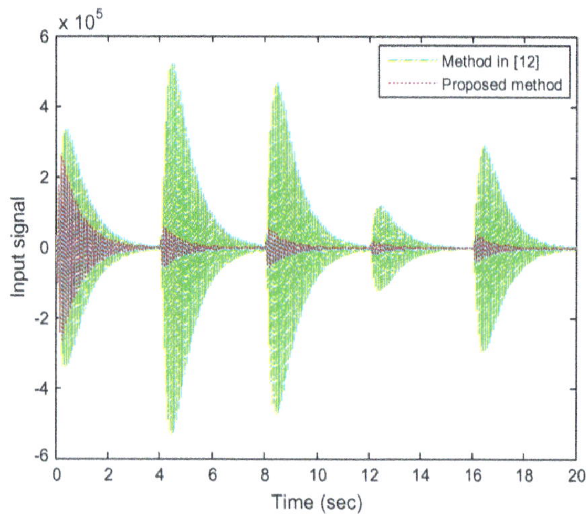

Fig. 2.9 The trajectory of control input

$$D(k) = 0.3\pi[\cos(3kT) + \sin(2kT)] \tag{2.63}$$

The network-induced disturbance $y_d(k)$ is:

$$y_d(k) = 4.5\sin(2kT) + 3.5\cos(2kT) \tag{2.64}$$

The gain of the observer is $\widehat{L} = [-0.1\,0\,0.3]^T$. The simulation time is 20 s.

Figures 2.10 and 2.11 show the trajectory of the external disturbance and network-induced disturbance, respectively. Figure 2.12 plots the trajectory of the tracking output for the vehicle velocity. The method in [12] produces noticeable jitter at the beginning, whereas the proposed method maintains stable tracking. Figure 2.13 gives

Fig. 2.10 The trajectory of external disturbance

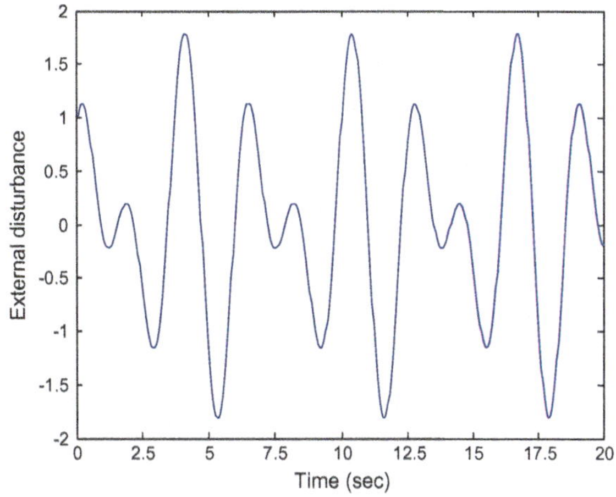

Fig. 2.11 The trajectory of network-induced disturbance

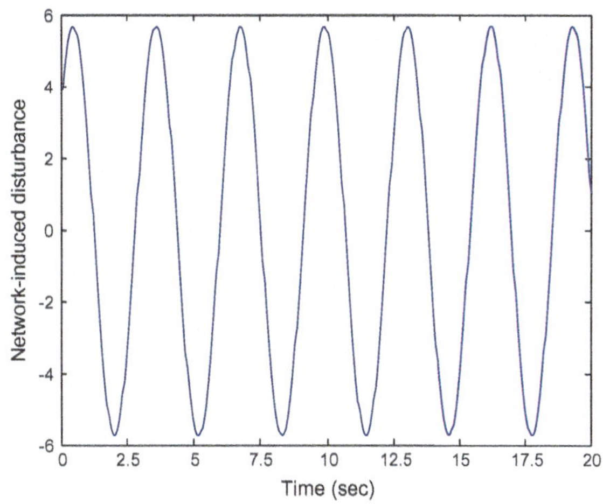

Fig. 2.12 The trajectory of tracking output for the velocity of the vehicle

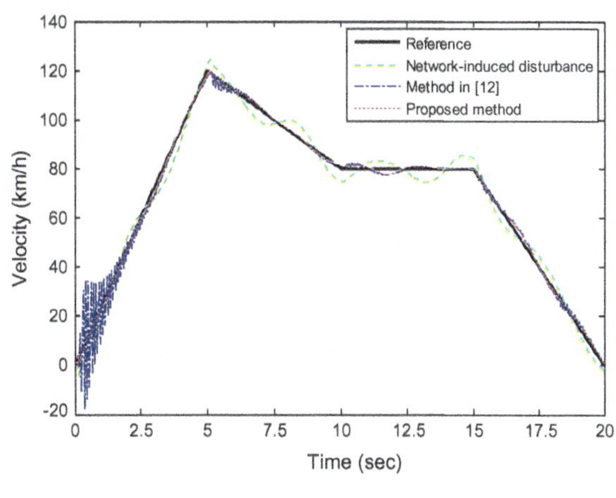

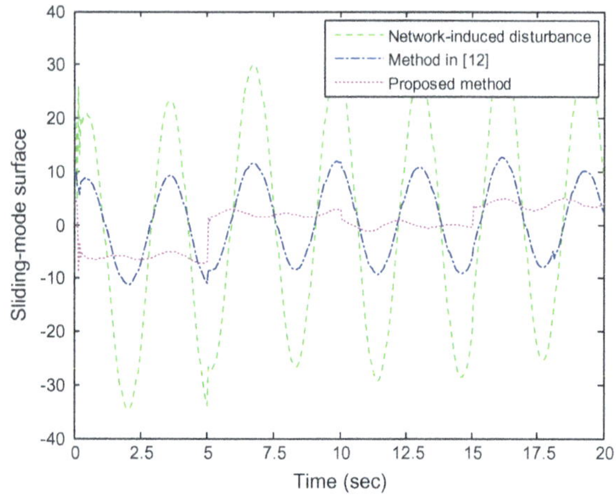

Fig. 2.13 The trajectory of the sliding-mode surface

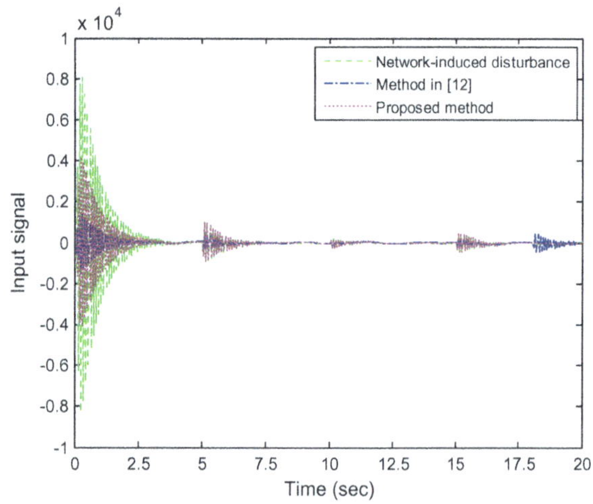

Fig. 2.14 The trajectory of the input signal

the trajectory of the sliding mode surface, and Fig. 2.14 presents the trajectory of the control input.

2.6 Conclusions

In this chapter, we discussed the tracking control of a networked intelligent vehicle with network-induced disturbance and external disturbance. A novel high-order adaptive discrete-time sliding-mode control strategy was proposed to suppress the external disturbance. The values of the estimated disturbance were used as adaptive factors, and an adaptive high-order sliding mode function was proposed. The sliding mode controller was designed using the equivalent control method, and the stability of the proposed method was validated. We also considered networked-induced disturbance in the vehicle network and proposed a state observed to compensate for this disturbance. Finally, the tracking output error was determined, and simulations of the networked vehicle were conducted to illustrate the effectiveness of the proposed controller and observer.

References

1. M. Li, Y. Chen, A. Zhou, W. He, X. Li, Adaptive tracking control for networked control systems of intelligent vehicle, Information Sciences, **503**, 493-507(2019)
2. Y. Lin, I. Rubin, Integrated message dissemination and traffic regulation for autonomous VANETs, IEEE Trans. Veh. Technol. **66** (10), 8644–8658 (2017)
3. S. Wen, G. Guo, B. Chen, X. Gao, Event-triggered cooperative control of vehicle platoons in vehicular ad hoc networks, Inf. Sci. (Ny). **459**, 341–353 (2018)
4. T. Nguyen, W. Su, Z. Gajic, C. Edwards, Higher accuracy output feedback sliding mode control of sampled-data systems, IEEE Trans. Automat. Contr. **61** (10), 3177–3182, (2016)
5. B. Zhang, Q. Han, X. Zhang, X. Yu, Sliding mode control with mixed current and delayed states for offshore steel jacket platforms, IEEE Trans. Contr. Syst. Technol. **22**(5), 1769–1783(2014)
6. J. Liu, S. Vazquez, L. Wu, A. Marquez, H. Gao, L. Franquelo, Extended state observer-based sliding-mode control for three-phase power converters, IEEE Trans. Indust. Electron. **64** (1) 22–31 (2017)
7. S. Biricik, H. Komurcugil, Optimized sliding mode control to maximize existence region for single-phase dynamic voltage restorers, IEEE Trans. Ind. Inf. **12**(4), 1486–1497(2016)
8. Č. Milosavljević, B. Draženović, B. Veselić, Discrete-time velocity servo system design using sliding mode control approach with disturbance compensation, IEEE Trans. Ind. Inf. **9**(2), 920–927 (2013)
9. M. Li, Y. Chen, Robust adaptive sliding mode control for switched networked control systems with disturbance and faults, IEEE Trans. Ind. Inf. **15**(1), 193–204(2019)
10. S. Ding, S. Li, Second-order sliding mode controller design subject to mismatched term, Automatica **77**, 388–392 (2017)
11. H. Du, X. Yu, M. Chen, S. Li, Chattering-free discrete-time sliding mode control, Automatica **68**, 87–91(2016)
12. Q. Xu, Piezoelectric nanopositioning control using second-order discrete-time terminal sliding-mode strategy, IEEE Trans. Indust. Electron. **62** (12), 7738–7748 (2015)
13. H. Castaneda, F. Plestan, A. Chriette, J. Morales, Continuous differentiator based on adaptive second-order sliding-mode control for a 3-dof helicopter, IEEE Trans. Indust. Electron. **63**(9), 5786–5793(2016)
14. C. Evangelista, A. Pisano, P. Puleston, E. Usai, Receding horizon adaptive second-order sliding mode control for doubly-fed induction generator based wind turbine, IEEE Trans. Control Syst. Technol. **25** (1), 73–84 (2017)
15. Y. Zhao, J. Wang, F. Yan, Y. Shen, Adaptive sliding mode fault-tolerant control for type-2 fuzzy systems with distributed delays, Inf. Sci. (Ny) **473**, 227–238(2019)
16. X. Ge, Q. Han, X. Zhang, Achieving cluster formation of multi-agent systems under aperiodic sampling and communication delays, IEEE Trans. Indust. Electron. **65** (4), 3417–3426(2018)
17. X. Ge, Q. Han, Consensus of multiagent systems subject to partially accessible and overlapping Markovian network topologies, IEEE Trans. Cybern. **47** (8), 1807–1819(2017)
18. H. Zhang, Y. Shi, M. Liu, $H\infty$ step tracking control for networked discrete-time nonlinear systems with integral and predictive actions, IEEE Trans. Ind. Inf. **9** (1), 337-345 (2013)

19. J. Geromel, G. Gabriel, Optimal H2 state feedback sampled-data control design of Markov jump linear systems, Automatica **54**, 182–188(2015)
20. D. Huang, S. Guang, State feedback control of uncertain networked control systems with random time delays, IEEE Trans. Automat. Contr. **53**(3), 829–834(2008)
21. G. Liu, Y. Xia, J. Chen, D. Rees, W. Hu, Networked predictive control of systems with random network delays in both forward and feedback channels, IEEE Trans. Indust. Electron. **54** (3), 1282–1297 (2007)
22. G. Liu, Consensus and stability analysis of networked multiagent predictive control systems, IEEE Trans. Cybern. **47**(4), 1114–1119(2017)
23. Y. Xia, W. Xie, B. Liu, X. Wang, Data-driven predictive control for networked control systems, Inf. Sci. (Ny) **235**, 45–54(2013)
24. J. Zhang, Y. Lin, P. Shi, Output tracking control of networked control systems via delay compensation controllers, Automatica **57**, 85–92(2015)
25. L. Zou, Z. Wang, H. Gao, Observer-based H ∞ control of networked systems with stochastic communication protocol: the finite-horizon case, Automatica **63**, 366-373 (2016)
26. Y. Alma, A. Nancy, L. Carlos, Bacterial foraging optimization algorithm to improve a discrete-time neural second order sliding mode controller, *Applied Mathematics and Computation*. **271**(15) 43-51(2015)
27. D. Ginoya, P. Shendge, S. Phadke, Delta-Operator-Based Extended Disturbance Observer and Its Applications, *IEEE Transactions on Industrial Electronics*. **62**(9) 5817-5828(2015)
28. Q. Xu, Piezoelectric nanopositioning control using second-order discrete-time terminal sliding-mode strategy, IEEE Transactions on Industrial Electronics. **62**(12) 7738-7748 (2015)
29. G. Guo, W. Yue, Autonomous platoon control allowing range-limited sensors, *IEEE Transactions on Vehicular Technology*. **61**(7) 2901-2912 (2012)

Chapter 3
Robust Sliding-Mode Tracking Control for NCSs with Random Delay and Packet Dropouts

In this chapter, the tracking control of networked control systems (NCSs) with packet dropouts and random delay in the feedforward channel (controller to actuator) and feedback channel (sensor to controller) is considered, and a networked DC permanent magnet motor (DC-PMM) system is investigated. A robust sliding mode tracking control method based on the pseudo partial derivative (PPD) theory is proposed [1] to compensate for packet dropouts and random delay in the network. The convergence of the tracking error is evaluated, and the stability of the proposed algorithm is discussed. Finally, a simulation of a DC-PMM in an NCS is carried out to demonstrate the performance of the proposed approach.

3.1 Introduction

The components of an NCS include the actuator, controller and sensor, which are connected by a wired or wireless network. Recently, NCSs have been widely used in various fields, for instance, teleoperation systems [2, 3] and cloud control systems [4]. Unlike a general control system, NCSs are characterized by resource sharing, simple installation, and easy maintenance, and they are easily extensible and have low cost and high reliability [5]. However, NCSs also have some challenges, such as jitter, packet dropouts, constraints on communication bandwidth, and time delay [6]. Delay and packet loss have the most significant effect on the control performance.

Many control algorithms have been developed to deal with these problems (e.g., [7, 8] and references therein). Network predictive control (NPC) is a typical control algorithm that actively compensates for delay and packet loss. We provide examples of studies on NPC. In [9], a data-based NPC method based was presented to compensate for data packet loss in two channels,. In [10], an observer-based NPC was designed to compensate for packet dropouts and distributed delays in the feedback channels. Reference [11] proposed a new NPC algorithm consisting of a network delay compensator and prediction generator. The former mitigated data dropout and network transmission delay, and the latter generated a set of future control

© The Author(s), under exclusive license to Springer Nature Singapore Pte Ltd. 2022
M. Li et al., *Tracking Control of Networked Systems via Sliding-Mode*,
https://doi.org/10.1007/978-981-16-6514-1_3

inputs to ensure that the closed-loop system reached the desired control target. In [12], a dynamic predictive feedback linearization control strategy was presented to compensate for data dropouts and networked communication delays to ensure stable system dynamics. A new NPC approach was designed in [13] to prevent data dropout and network delays. A state-based NPC approach with packet-based transmission was proposed in [14] to compensate for network communication delays. However, the aforementioned NPC algorithms only considered packet dropouts or delays or assumed that the feedback channel or feedforward channel was ideal. Several strategies were proposed to provide an adequate uncertainty description and an accurate model of the plant.

Recently, PMMs have attracted significant attention as a high-performance electric drive system due to their considerable advantages, such as high power density, large range of operating speed, low maintenance, and a large torque to inertia ratio [15]. Many control algorithms have been reported for DC motors in the literature (see, e.g., [16–20] and references therein). In traditional DC motor control, the system components are connected by point-to-point wiring and are located in the same place.

However, in practical applications, such as aeronautics, astronautics, and teleoperation systems, the controlled object and controller are located in different places, and the control signal needs to be transmitted from one location to another. Therefore, these system components are usually connected by communication networks, i.e., the NCS. The control of a DC motor in NCSs is affected by packet dropouts and network-induced delay, resulting in low control performance. In [19], a tracking output controller with two control parameters was proposed, and the packet losses and time delays packet losses and delays were regarded as parameter perturbations. The problem of designing the controller was transformed into an optimization problem. However, the accurate motor model need know when design of the controller.

3.2 Sliding Mode Control for DC-PMM

3.2.1 *Model of DC-PMM*

The DC motor system can be modeled as follows [20]:

$$\overline{L}_a \frac{d\overline{i}_a}{dt} = \overline{\upsilon} - \overline{R}_a \overline{i}_a - \overline{k}_e \overline{\omega} \tag{3.1}$$

$$\overline{J}\frac{d\overline{\omega}}{dt} = -b'\overline{\omega} + \overline{k}_m \overline{i}_a \tag{3.2}$$

where $\overline{\omega}$, $\overline{k}_e$, $\overline{\upsilon}$, $\overline{L}_a$, $\overline{i}_a$, and $\overline{k}_m$ are the angular velocity, counter-electromotive force constant, input voltage in the motor armature terminals, armature inductance, armature current, and motor torque constant, respectively. $\overline{J}$ represents the moment of inertia of the rotor and motor load, $\overline{R}_a$ describes the armature resistance, and b' is the viscous friction coefficient of the motor.

For the analysis, we define:

$$\overline{F} = \overline{\omega} \tag{3.3}$$

Hence, the differential parameterization of the system variables is given by the following equation:

$$\overline{i}_a = \frac{1}{\overline{k}_m}(\overline{J\dot{F}} + b'\overline{F}) \tag{3.4}$$

$$\overline{\omega} = \overline{F} \tag{3.5}$$

$$\frac{\overline{JL_a}}{\overline{k}_m}\ddot{\overline{F}} + \frac{1}{\overline{k}_m}(b'\overline{L}_a + \overline{JR_a})\dot{\overline{F}} + (\frac{b'\overline{R}_a}{\overline{k}_m} + \overline{k}_e)\overline{F} = \overline{\upsilon} \tag{3.6}$$

Equation (3.6) can be formulated as follows:

$$\overline{a}_0\ddot{\overline{F}} + \overline{a}_1\dot{\overline{F}} + \overline{a}_2\overline{F} = \overline{\upsilon} \tag{3.7}$$

where $\overline{a}_0 = \frac{\overline{JL_a}}{\overline{k}_m}$, $\overline{a}_1 = \frac{1}{\overline{k}_m}(b'\overline{L}_a + \overline{JR_a})$, $\overline{a}_2 = (\frac{b'\overline{R}_a}{\overline{k}_m} + \overline{k}_e)$.

Based on the above equations, the model can be discretized as:

$$\overline{a}_0\ddot{\overline{F}}(k) + \overline{a}_1\dot{\overline{F}}(k) + \overline{a}_2\overline{F}(k) = \overline{\upsilon}(k) \tag{3.8}$$

Let $\overline{y}(k) = \dot{\overline{F}}(k)$ and $u(k) = \overline{\upsilon}(k)$ to obtain:

$$\dot{\overline{y}}(k) = \frac{u(k)}{\overline{a}_0} - \frac{\overline{a}_1}{\overline{a}_0}\overline{y}(k) - \frac{\overline{a}_2}{\overline{a}_0}\overline{G}(k) \tag{3.9}$$

where $\overline{G}(k) = \int \overline{y}(t)dt$.

We substitute $\dot{\overline{y}}(k) = \frac{\overline{y}(k+1)-\overline{y}(k)}{T}$ into Eq. (3.9). The kth moment discrete-time form of this model can be represented as follows, which is the transformed form of the DC-PMM dynamics:

$$\overline{y}(k+1) = (1 - \frac{T\overline{a}_1}{\overline{a}_0})\overline{y}(k) + \frac{T}{\overline{a}_0}u(k) + \overline{g}(k) \tag{3.10}$$

where $\overline{g}(k)$ describes the noise and external disturbances, and $\overline{g}(k) = -\frac{T\overline{a}_2}{\overline{a}_0}\overline{G}(k)$.

The partial derivatives of $\overline{y}(k+1)$ in Eq. (3.10) of the control inputs $u(k)$,$u(k-1)$, $\cdots$, and the outputs $\overline{y}(k)$, $\overline{y}(k-1)$, $\cdots$ are continuous, and the system is a generalized

Lipschitz function, i.e., $|\Delta\overline{y}(k+1)| \leq b||\Delta \mathrm{U(k)}||$ for $\forall k$ and $||\Delta \mathrm{U(k)}|| \neq 0$, where $\Delta\overline{y}(k+1) = \overline{y}(k+1) - \overline{y}(k)$, $\Delta \mathrm{U(k)} = [\Delta \mathrm{u(k)}, \Delta \mathrm{u(k-1)}, \ldots, \Delta u(k-L+1)]^T$, $||\cdot||$ denotes 2-norm, b is a constant, and L is the length of the control input.

For a nonlinear system (3.10), if L is given, there must exist $\overline{\phi}(k)$ and $||\overline{\phi}(k)|| \leq b$, which is satisfied if $||\overline{\phi}(k)|| \neq 0$. The system can be transformed into the following partial form of the equivalent dynamic linearization data model [7]:

$$\Delta\overline{y}(k+1) = \overline{\phi}^T(k)\Delta \mathrm{U(k)} \tag{3.11}$$

where $\overline{\phi}(k) = [\overline{\phi}_1(k), \overline{\phi}_2(k), \ldots, \overline{\phi}_L(k)]^T$, and $\overline{\phi}(k)$ represents the PPD. The equation of the system (3.11) can be rewritten as follows:

$$\overline{X}(k+1) = \overline{A}\overline{X}(k) + \overline{B}\Delta u(k) \tag{3.12}$$

$$\Delta\overline{y}(k) = \overline{\mathrm{C}}^T(k)\overline{X}(k) \tag{3.13}$$

where $\overline{X}(k) = [\Delta u(k-1), \Delta u(k-2), \ldots, \Delta u(k-L)]^T$,

$$\overline{A} = \begin{pmatrix} 0 & 0 & \cdots & 0\,0 \\ 1 & 0 & \cdots & 0\,0 \\ 0 & 1 & \cdots & 0\,0 \\ \vdots & \vdots & \cdots & 0\,0 \\ 0 & 0 & \cdots & 1\,0 \end{pmatrix}$$

$$\overline{B} = [1\,0\,0\,\cdots 0\,0\,]^T$$

$$\overline{C}(k) = [\overline{\phi}_1(k-1), \overline{\phi}_2(k-2), \ldots, \overline{\phi}_L(k-L)]^T$$

$$||\overline{C}(k)|| \leq b$$

Based on the PPD theory proposed in [7], the estimated value of $\overline{C}(k)$ can be obtained using the following method:

$$\begin{aligned} \hat{\bar{C}}(k) = \hat{\bar{C}}(k-1) + &\frac{\bar{\alpha}\bar{X}(k-1)}{\bar{\beta} + \bar{X}^T(k-1)\bar{X}(k-1)} \\ &\times [\bar{y}(k) - \bar{y}(k-1) - \bar{X}^T(k-1)\hat{\bar{C}}(k-1)] \end{aligned} \tag{3.14}$$

where $\beta(>0)$ is the weight factor, $\hat{C}(k) = [\hat{\phi}_1(k-1), \hat{\phi}_2(k-2), \ldots, \hat{\phi}_L(k-L)]^T$, $\overline{\alpha}$ is the step factor, and $\overline{\alpha} \in (0, 1]$.

3.2.2 Sliding Mode Control for DC-PMM

According to Eqs. (3.12) and (3.13), the tracking error can be defined as:

$$\overline{e}(k) = \overline{y}_d(k) - \overline{y}(k) \tag{3.15}$$

where $\overline{y}_d(k)$ is the reference output signal.

We design the sliding surface function as follows:

$$\overline{s}(k) = \overline{\lambda}\overline{e}(k) + \overline{\gamma}\Delta\overline{e}(k) \tag{3.16}$$

where $\overline{\lambda} > 0, \overline{\gamma} > 0$ are weight factors, and $\Delta\overline{e}(k) = \overline{e}(k) - \overline{e}(k-1)$.

Hence, by combining Eqs. (3.16) and (3.13), we obtain:

$$\begin{aligned}
&\overline{s}(k+1)\\
&= \overline{\lambda}\overline{e}(k+1) + \overline{\gamma}\Delta\overline{e}(k+1)\\
&= (\overline{\lambda} + \overline{\gamma})\overline{e}(k+1) - \overline{\gamma}\overline{e}(k)\\
&= (\overline{\lambda} + \overline{\gamma})(\overline{y}_d(k+1) - \overline{y}(k+1)) - \overline{\gamma}(\overline{y}_d(k) - \overline{y}(k))\\
&= (\overline{\lambda} + \overline{\gamma})(\overline{y}_d(k+1) - \overline{y}(k) - \overline{C}^T(k+1)\overline{X}(k+1)) - \overline{\gamma}(\overline{y}_d(k) - \overline{y}(k))
\end{aligned} \tag{3.17}$$

We select the sliding motion reaching law as follows [21]:

$$\overline{s}(k+1) = (1 - \overline{q}T)\overline{s}(k) - \overline{\varepsilon}T sgn(\overline{s}(k)) \tag{3.18}$$

where $\overline{\varepsilon} > 0, \overline{q} > 0, 1 - \overline{q}T > 0$, $sgn(\cdot)$ is a symbolic function.

We substitute Eq. (3.18) into Eq. (3.17):

$$\begin{aligned}
&\overline{C}^T(k+1)\overline{\mathbf{x}}(k+1)\\
&= (\overline{y}_d(k+1) - \overline{y}(k)) - (\overline{\lambda} + \overline{\gamma})^{-1}\\
&[(1 - \overline{q}T)\overline{s}(k) - \overline{\varepsilon}T sgn(\overline{s}(k)) + \overline{\gamma}(\overline{y}_d(k) - \overline{y}(k))]
\end{aligned} \tag{3.19}$$

We substitute Eqs. (3.12) into (3.19):

$$\begin{aligned}
&\Delta u(k)\\
&= [\overline{C}^T(k+1)\overline{B}]^{-1}\{(\overline{y}_d(k+1) - \overline{y}(k)) - (\overline{\lambda} + \overline{\gamma})^{-1}[(1 - \overline{q}T)\\
&\times\overline{s}(k) - \overline{\varepsilon}T sgn(\overline{s}(k)) + \overline{\gamma}(\overline{y}_d(k) - \overline{y}(k))] - \overline{C}^T(k+1)\overline{A}\overline{X}(k)\}
\end{aligned} \tag{3.20}$$

3.2.3 Stability Analysis

Two theorems are proposed to analyze the stability of the sliding motion:

Theorem 1 *For discrete-time systems* (3.12) *and* (3.13), *when the reaching law is defined as Eqs.* (3.18), *the control strategy of the system is stable, and any initial state will be close to the sliding surface.*

Proof The proof steps have been omitted due to the similar procedures in [22].

Theorem 2 *For a nonlinear system* (3.10), *when* $\overline{y}_d(k) = ref = constant$ *and* $\overline{\lambda} > 0$, *the closed-loop system ensures that*:

$$\lim_{k\to\infty}(ref - \overline{y}(k)) = 0 \tag{3.21}$$

Proof Let $\overline{e}(k) = ref - \overline{y}(k)$; according to Eqs. (3.12), (3.13), and (3.20), the following applies:

$$\begin{aligned}
&\overline{e}(k+1)\\
&= ref - \overline{y}(k+1) = (ref - \overline{y}(k)) - \overline{C}^T(k+1)\overline{X}(k+1)\\
&= \overline{e}(k) - \overline{C}^T(k+1)[\overline{AX}(k) + \overline{B}\Delta u(k)]\\
&= \overline{e}\lim_{x\to\infty}(k) - \overline{C}^T(k+1)\overline{AX}(k) - \overline{C}^T(k+1)\overline{B}[\overline{C}^T(k+1)\overline{B}]^{-1}\\
&\quad\times\{(\overline{y}_d(k+1) - \overline{y}(k)) - (\overline{\lambda}+\overline{\gamma})^{-1}[(1-\overline{q}T)\overline{s}(k) - \overline{\varepsilon}T sgn(\overline{s}(k))\\
&\quad+\overline{\gamma}(\overline{y}_d(k) - \overline{y}(k))] - \overline{C}^T(k+1)\overline{AX}(k)\}\\
&= \overline{e}(k) - \overline{C}^T(k+1)\overline{AX}(k) - \{\overline{e}(k) - (\overline{\lambda}+\overline{\gamma})^{-1}[(1-\overline{q}T)\overline{s}(k)\\
&\quad-\overline{\varepsilon}T sgn(\overline{s}(k)) + \overline{\gamma}\overline{e}(k)] - \overline{C}^T(k+1)\overline{AX}(k)\}\\
&= (\overline{\lambda}+\overline{\gamma})^{-1}[(1-\overline{q}T)\overline{s}(k) - \overline{\varepsilon}T sgn(\overline{s}(k)) + \overline{\gamma}\overline{e}(k)]\\
&= \frac{\overline{\gamma}}{\overline{\lambda}+\overline{\gamma}}\overline{e}(k) + \frac{(1-\overline{q}T)\overline{s}(k) - \overline{\varepsilon}T sgn(\overline{s}(k))}{\overline{\lambda}+\overline{\gamma}}
\end{aligned} \tag{3.22}$$

Because $\overline{\lambda} > 0, \overline{\gamma} > 0$ and $\lim_{k\to\infty}\overline{s}(k) = 0$, thus $\frac{\overline{\gamma}}{\overline{\lambda}+\overline{\gamma}} \in (0, 1)$, and $\lim_{k\to\infty}\overline{e}(k) = 0$. This completes the proof.

3.3 Sliding Mode Predictive Tracking Control Based on PPD for a DC-PMM in NCSs

As shown in Fig. 3.1, the control structure of the DC-PMM in NCSs is composed of driving motors, sensors, parameter estimators, controllers, predictive generators,

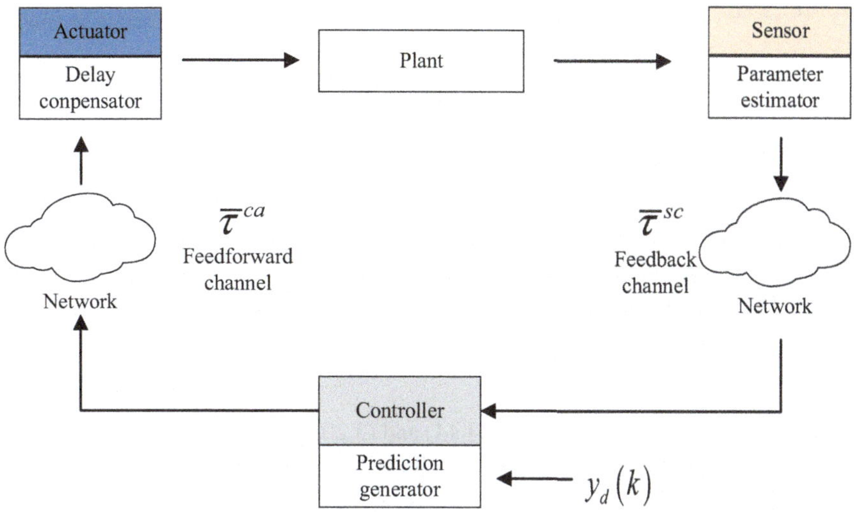

Fig. 3.1 Structure diagram of control of DC-PMM in NCSs

delay compensators, and actuators. The delays in the feedback channel (sensor-controller) and feedforward channel (controller-actuator) are $\overline{\tau}^{sc}(k)$ and $\overline{\tau}^{ca}(k)$, respectively. The system satisfies the following assumptions.

Assumption 1 The sensor is driven by time, and the controller and actuator are driven by events. Each node is clock-synchronized.

Assumption 2 $\overline{\tau}^{sc}(k)$ and $\overline{\tau}^{ca}(k)$ are bounded, i.e., $0 \le \overline{\tau}^{sc}(k) \le \tilde{\overline{\tau}}^{sc}$, $0 \le \overline{\tau}^{ca}(k) \le \tilde{\overline{\tau}}^{ca}$, where $\tilde{\overline{\tau}}^{sc}$, $\tilde{\overline{\tau}}^{ca}$ are constant.

Remark 1 The data packets are time-stamped when they are sent from the sensors and controllers, and data transmission occurs in the form of a single data packet. The controller and actuator always select the latest data and discard the old data packets. When the old data packets arrive, they are treated as lost data packets. Data loss can be regarded as a special network communication delay and is combined into $\overline{\tau}^{sc}(k)$ or $\overline{\tau}^{ca}(k)$. For example, if the number of consecutive packet losses is X, the delay caused by the packet loss is $\overline{\tau}_d = XT$, and $\overline{\tau}_d \le \tilde{\overline{\tau}}_d$, where $\tilde{\overline{\tau}}_d$ is a constant.

3.3.1 Sensor

In sensor, Eqs. (3.14) and (3.20) are utilized to obtain $\hat{\overline{C}}(k)$. Packing $\overline{Y}(k) = (\overline{y}(k), \overline{y}(k-1), ..., \overline{y}(k - \tilde{\overline{\tau}}^{sc} - \tilde{\overline{\tau}}_d))$,$\overline{\psi}(k) = (\hat{\overline{C}}(k), \hat{\overline{C}}(k-1), ..., \hat{\overline{C}}(k - \tilde{\overline{\tau}}^{sc} - \tilde{\overline{\tau}}_d))$,$U'(k) = (u(k-1), u(k-2), \ldots, u(k - \tilde{\overline{\tau}}^{sc} - \tilde{\overline{\tau}}_d))$ together with time-stamp, then send them to controller. That is

$$\overline{P}_k = [\overline{Y}(k), \overline{\psi}(k), U'(k)] \tag{3.23}$$

3.3.2 Controller

We set the prediction generator in the controller and assume that the current moment is k. $\overline{\tau}^{sc}(k)$ can be obtained by comparing the local time with the time stamp in the packet. The feedback information received by the controller is as follows:

$$\overline{P}_{k-\overline{\tau}^{sc}(k)} = [\overline{Y}(k-\overline{\tau}^{sc}(k)), \overline{\psi}(k-\overline{\tau}^{sc}(k)), U'(k-\overline{\tau}^{sc}(k))]^T \tag{3.24}$$

According to Eqs. (3.12), (3.13), (3.14) and (3.20), we use the received feedback information to predict the next moment information as follows:

$$\begin{aligned}
&\Delta\overline{y}(k-\overline{\tau}^{sc}(k)+1|k-\overline{\tau}^{sc}(k)) \\
&= \hat{\overline{C}}^T(k-\overline{\tau}^{sc}(k))\overline{X}(k-\overline{\tau}^{sc}(k)+1|k-\overline{\tau}^{sc}(k)) \\
&= \hat{\overline{C}}^T(k-\overline{\tau}^{sc}(k))[\overline{AX}(k-\overline{\tau}^{sc}(k)) + \overline{B}\Delta u(k-\overline{\tau}^{sc}(k))]
\end{aligned} \tag{3.25}$$

$$\begin{aligned}
&\overline{y}(k-\overline{\tau}^{sc}(k)+1|k-\overline{\tau}^{sc}(k)) \\
&= \overline{y}(k-\overline{\tau}^{sc}(k)) + \Delta\overline{y}(k-\overline{\tau}^{sc}(k)+1|k-\overline{\tau}^{sc}(k))
\end{aligned} \tag{3.26}$$

$$\begin{aligned}
&\overline{s}(k-\overline{\tau}^{sc}(k)+1|k-\overline{\tau}^{sc}(k)) \\
&= (\overline{\lambda}+\overline{\gamma})(\overline{y}_d(k-\overline{\tau}^{sc}(k)+1) - \overline{y}(k-\overline{\tau}^{sc}(k)+1|k-\overline{\tau}^{sc}(k))) \\
&\quad - \overline{\gamma}(\overline{y}_d(k-\overline{\tau}^{sc}(k)) - \overline{y}(k-\overline{\tau}^{sc}(k)))
\end{aligned} \tag{3.27}$$

$$\begin{aligned}
&\Delta u(k-\overline{\tau}^{sc}(k)+1|k-\overline{\tau}^{sc}(k)) \\
&= [\hat{\overline{C}}^T(k-\overline{\tau}^{sc}(k))\overline{B}]^{-1}\{(\overline{y}_d(k-\overline{\tau}^{sc}(k)+2) \\
&\quad - \overline{y}(k-\overline{\tau}^{sc}(k)+1|k-\tau^{sc}(k))) - (\overline{\lambda}+\overline{\gamma})^{-1}[(1-\overline{q}T) \\
&\quad \times \overline{s}(k-\overline{\tau}^{sc}(k)+1|k-\tau^{sc}(k)) - \overline{\varepsilon}T sgn(\overline{s}(k-\overline{\tau}^{sc}(k)+1|k-\overline{\tau}^{sc}(k))) \\
&\quad + \overline{\gamma}(\overline{y}_d(k-\overline{\tau}^{sc}(k)+1) - \overline{y}(k-\overline{\tau}^{sc}(k)+1|k-\overline{\tau}^{sc}(k)))] \\
&\quad - \hat{\overline{C}}^T(k-\overline{\tau}^{sc}(k))[\overline{A}^2\overline{x}(k-\overline{\tau}^{sc}(k)) + \overline{AB}u(k-\overline{\tau}^{sc}(k))]\}
\end{aligned} \tag{3.28}$$

$$\begin{aligned}
&u(k-\overline{\tau}^{sc}(k)+1|k-\overline{\tau}^{sc}(k)) \\
&= u(k-\overline{\tau}^{sc}(k)) + \Delta u(k-\overline{\tau}^{sc}(k)+1|k-\overline{\tau}^{sc}(k))
\end{aligned} \tag{3.29}$$

In turn, we predict the control input of the system from moment $k-\overline{\tau}^{sc}(k)+2$ to $k-\overline{\tau}^{sc}(k)+\tilde{\overline{\tau}}^{sc}$.

$$\begin{aligned}&\Delta\overline{y}(k-\overline{\tau}^{sc}(k)+i|k-\overline{\tau}^{sc}(k))\\&=\hat{\overline{C}}^{T}(k-\overline{\tau}^{sc}(k))\overline{X}(k-\overline{\tau}^{sc}(k)+i|k-\overline{\tau}^{sc}(k))\\&=\hat{\overline{C}}^{T}(k-\overline{\tau}^{sc}(k))(\overline{A}^{i}\overline{x}(k-\overline{\tau}^{sc}(k))+\Delta u(k-\overline{\tau}^{sc}(k))\sum_{l=1}^{i-1}\overline{A}^{l}\overline{B}\\&\quad+\overline{B}\Delta u(k-\overline{\tau}^{sc}(k)+(i-1)))\end{aligned}\tag{3.30}$$

$$\begin{aligned}&\overline{y}(k-\overline{\tau}^{sc}(k)+i|k-\overline{\tau}^{sc}(k))\\&=\overline{y}(k-\overline{\tau}^{sc}(k)+(i-1)|k-\overline{\tau}^{sc}(k))+\Delta\overline{y}(k-\overline{\tau}^{sc}(k)+i|k-\overline{\tau}^{sc}(k))\end{aligned}\tag{3.31}$$

$$\begin{aligned}&\overline{s}(k-\overline{\tau}^{sc}(k)+i|k-\overline{\tau}^{sc}(k))\\&=(\overline{\lambda}+\overline{\gamma})(\overline{y}_d(k-\overline{\tau}^{sc}(k)+i)\\&\quad-\overline{y}(k-\overline{\tau}^{sc}(k)+i|k-\overline{\tau}^{sc}(k)))\\&\quad-\overline{y}(\overline{y}_d(k-\overline{\tau}^{sc}(k)+(i-1))\\&\quad-\overline{y}(k-\overline{\tau}^{sc}(k)+(i-1)|k-\overline{\tau}^{sc}(k)))\end{aligned}\tag{3.32}$$

$$\begin{aligned}&\Delta u(k-\overline{\tau}^{sc}(k)+i|k-\overline{\tau}^{sc}(k))\\&=[\hat{\overline{C}}^{T}(k-\overline{\tau}^{sc}(k))\overline{B}]^{-1}\{(\overline{y}_d(k-\overline{\tau}^{sc}(k)+(i+1))\\&\quad-\overline{y}(k-\overline{\tau}^{sc}(k)+i|k-\overline{\tau}^{sc}(k)))-(\overline{\lambda}+\overline{\gamma})^{-1}[(1-\overline{q}T)\overline{s}(k)\\&\quad-\overline{\tau}^{sc}(k)+i|k-\overline{\tau}^{sc}(k))-\overline{\varepsilon}T\,sgn(\overline{s}(k-\overline{\tau}^{sc}(k)+i|k-\overline{\tau}^{sc}(k)))\\&\quad+\overline{\gamma}(\overline{y}_d(k-\overline{\tau}^{sc}(k)+i)-\overline{y}(k-\overline{\tau}^{sc}(k)+i|k-\overline{\tau}^{sc}(k)))]\\&\quad-\hat{\overline{C}}^{T}(k-\overline{\tau}^{sc}(k))[\overline{A}^{i+1}\overline{x}(k-\overline{\tau}^{sc}(k))+u(k-\overline{\tau}^{sc}(k))\sum_{l=1}^{i-1}\overline{A}^{l+1}\overline{B}\\&\quad+\overline{AB}\Delta u(k-\overline{\tau}^{sc}(k)+(i-1))]\}\end{aligned}\tag{3.33}$$

$$\begin{aligned}&u(k-\overline{\tau}^{sc}(k)+i|k-\overline{\tau}^{sc}(k))\\&=u(k-\overline{\tau}^{sc}(k)+(i-1)|k-\overline{\tau}^{sc}(k))+\Delta u(k-\overline{\tau}^{sc}(k)+i|k-\overline{\tau}^{sc}(k))\end{aligned}\tag{3.34}$$

For $i=2,\ldots,\tilde{\overline{\tau}}^{sc}$.

Thereby, the predicted value of the k instant control signal is obtained. In addition, we consider $\overline{\tau}^{ca}(k)$ in the feedforward channel, but the model information of the system is unknown. Therefore, the prediction generator must also predict the control input from $k+1$ to $k+\tilde{\overline{\tau}}^{ca}$, which is:

$$\begin{aligned}&\Delta\overline{y}(k+z|k-\overline{\tau}^{sc}(k))\\&=\hat{\overline{C}}^{T}(k-\overline{\tau}^{sc}(k))[\overline{A}^{z+\overline{\tau}^{sc}(k)}\overline{X}(k-\overline{\tau}^{sc}(k))+u(k-\overline{\tau}^{sc}(k))\\&\quad\times\sum_{l=1}^{z+\overline{\tau}^{sc}(k)-1}\overline{A}^{l}\overline{B}+\overline{B}\Delta u(k-\overline{\tau}^{sc}(k)+(z+\overline{\tau}^{sc}(k)-1))]\end{aligned}\tag{3.35}$$

$$\begin{aligned}&\overline{y}(k+z|k-\overline{\tau}^{sc}(k))\\&=\overline{y}(k+(z-1)|k-\overline{\tau}^{sc}(k))+\Delta\overline{y}(k+z|k-\overline{\tau}^{sc}(k))\end{aligned}\tag{3.36}$$

$$\begin{aligned}&\overline{s}(k+z|k-\overline{\tau}^{sc}(k))\\&=(\overline{\lambda}+\overline{\gamma})(\overline{y}_d(k+z)-\overline{y}(k+z|k-\overline{\tau}^{sc}(k)))\\&\quad-\overline{\gamma}(\overline{y}_d(k+z-1)-\overline{y}(k+z-1|k-\overline{\tau}^{sc}(k)))\end{aligned}\tag{3.37}$$

$$\begin{aligned}&\Delta u(k+z|k-\overline{\tau}^{sc}(k))\\&=[\hat{\overline{C}}^{T}(k-\overline{\tau}^{sc}(k))\overline{B}]^{-1}\{(\overline{y}_d(k-\overline{\tau}^{sc}(k)+(z+1))\\&\quad-\overline{y}(k+z|k-\overline{\tau}^{sc}(k)))-(\overline{\lambda}+\overline{\gamma})^{-1}[(1-\overline{q}T)\\&\overline{s}(k+z|k-\overline{\tau}^{sc}(k))-\overline{\varepsilon}T\,sgn(\overline{s}(k+z|k-\overline{\tau}^{sc}(k)))\\&\quad+\overline{\gamma}(\overline{y}_d(k+z)-\overline{y}(k+z|k-\overline{\tau}^{sc}(k)))]-\hat{\overline{C}}^{T}(k-\overline{\tau}^{sc}(k))\\&[\overline{A}^{z+\overline{\tau}^{sc}(k)+1}\overline{X}(k-\overline{\tau}^{sc}(k))+u(k-\overline{\tau}^{sc}(k))\sum_{l=1}^{z+\overline{\tau}^{sc}(k)-1}\overline{A}^{l+1}\overline{B}\\&\quad+\overline{AB}\Delta u(k-\overline{\tau}^{sc}(k)+(z+\overline{\tau}^{sc}(k)-1))]\}\end{aligned}\tag{3.38}$$

$$\begin{aligned}&u(k+z|k-\overline{\tau}^{sc}(k))\\&=u(k+(z-1)|k-\overline{\tau}^{sc}(k))+\Delta u(k+z|k-\overline{\tau}^{sc}(k))\end{aligned}\tag{3.39}$$

where $z=1,2,...,\tilde{\overline{\tau}}^{ca}$. We use $u(k+z|k-\overline{\tau}^{sc}(k))$ as the control sequence $U''_{k-\overline{\tau}^{sc}(k)}=[u_{k|k-\overline{\tau}^{sc}(k)},u_{k+1|k-\overline{\tau}^{sc}(k)},...,u_{k+\tilde{\overline{\tau}}^{ca}|k-\overline{\tau}^{sc}(k)}]$ and include the time stamp and send them to the actuator.

3.3.3 Actuator

The compensator is located on one side of the actuator; it stores the latest control sequence and compensates for $\overline{\tau}^{ca}(k)$ in the feedforward channel. First, the time stamp $\overline{\tau}^{ca}(k)$ in the data packet is compared and calculated with the local time; next, the controller generates the current control sequence $U''_{k-\overline{\tau}^{sc}(k)}$ received by the compensator. Since the delay of the feedback channel is $\overline{\tau}^{sc}(k-\overline{\tau}^{ca}(k))$, the delay of the entire circuit is $\overline{\tau}^{ca}(k)+\overline{\tau}^{sc}(k-\overline{\tau}^{ca}(k))$; thus, the output signal of the controlled system is $\overline{y}(k-\overline{\tau}^{ca}(k)-\overline{\tau}^{sc}(k-\overline{\tau}^{ca}(k)))$, and the input signal is $u(k-\overline{\tau}^{ca}(k)-\overline{\tau}^{sc}(k-\overline{\tau}^{ca}(k)))$. Therefore, at k moment, the control sequence of the compensator is:

$$\begin{aligned}U''_{k-\overline{\tau}^{ca}(k)-\overline{\tau}^{sc}(k-\overline{\tau}^{ca}(k))} &= [u_{k-\overline{\tau}^{ca}(k)|k-\overline{\tau}^{ca}(k)-\overline{\tau}^{sc}(k-\overline{\tau}^{ca}(k))},\\ &u_{k-\overline{\tau}^{ca}(k)+1|k-\overline{\tau}^{ca}(k)-\overline{\tau}^{sc}(k-\overline{\tau}^{ca}(k))}, \ldots,\\ &u_{k-\overline{\tau}^{ca}(k)+\overline{\overline{\tau}}^{ca}(k)|k-\overline{\tau}^{ca}(k)-\overline{\tau}^{sc}(k-\overline{\tau}^{ca}(k))}]\end{aligned} \tag{3.40}$$

The control input is:

$$u(k) = u_{k|k-\overline{\tau}^{ca}(k)-\overline{\tau}^{sc}(k-\overline{\tau}^{ca}(k))} \tag{3.41}$$

3.4 Results and Analysis

In this section, a DC/DC step-down power converter-DC motor system with a closed loop is considered. The motor control device is the same as that in [20] (GNM5440). The nominal voltage and power are 24 V and 95 W, respectively. The other parameters are $\overline{k}_e = 1.201 \times 10^{-1}(V \cdot s/rad)$, $\overline{L}_a = 2.22 \times 10^{-3}(H)$, $\overline{k}_m = 1.201 \times 10^{-1}(N \cdot m/A)$, $\overline{J} = 1.182 \times 10^{-1}(kgm^2)$, $\overline{Ra} = 0.965(\Omega)$, and $\overline{b'} = 1.296 \times 10^{-1}(N/ms)$.

Two examples are presented in this chapter to verify the performance of the proposed approach for a DC motor system in an NCS. Network-induced delays occur in both the feedback and forward channels.

The desired output is set as follows:

$$\overline{R}(k) = \begin{cases} 300, & 0 < k \le 101 \\ 450, & 101 < k \le 301 \\ 500, & 301 < k \le 500 \end{cases}$$

The initial conditions are $u(-2) = 2$, $u(-1) = 1.7$, $y(-1) = 20$,$y(0) = 30$, $\overline{\alpha} = 0.28$, $\overline{\beta} = 0.8$, and $\hat{\overline{\phi}}_1(0) = 20$. The parameters of the sliding mode controller are $\overline{\varepsilon} = 0.001$,$\overline{q} = 30$,$\overline{\lambda} = 0.99$, $\overline{\gamma} = 0.105$, and $T = 0.001$.

Case 1: constant delay in the feedback channel and feedforward channel. In this case, for all $k \in \{1, 2, \ldots, 500\}$, $\overline{\tau}^{sc}(k) = 1$,$\overline{\tau}^{ca}(k) = 1$.

The simulation results are shown in Figs. 3.2 and 3.3, depicting the input and output of the platform, respectively. As shown in Fig. 3.3, when there is a constant delay in the NCS, the output of the system exhibits significant jitter. After using the proposed method, the output of the DC motor can track the desired trajectory more quickly, and a stable output is obtained.

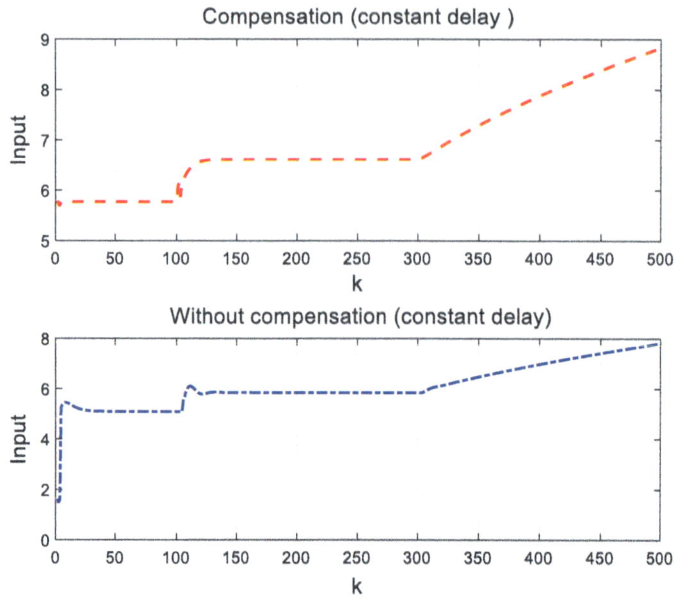

Fig. 3.2 Input of the plant with constant delay

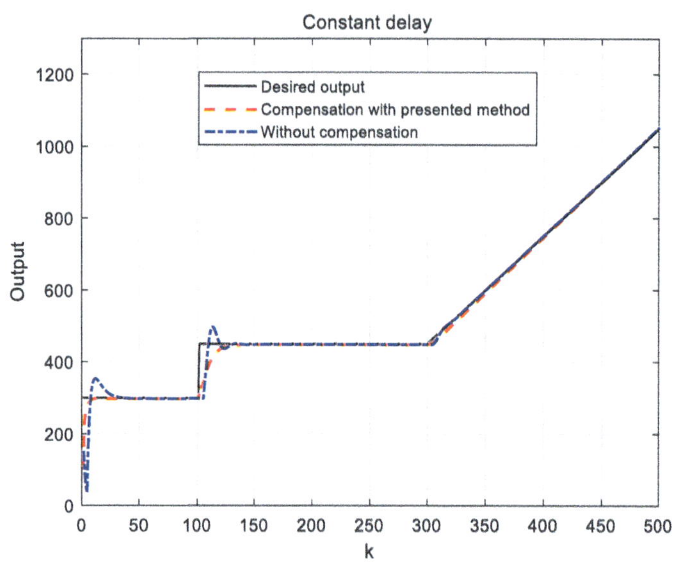

Fig. 3.3 Output of the plant with constant delay

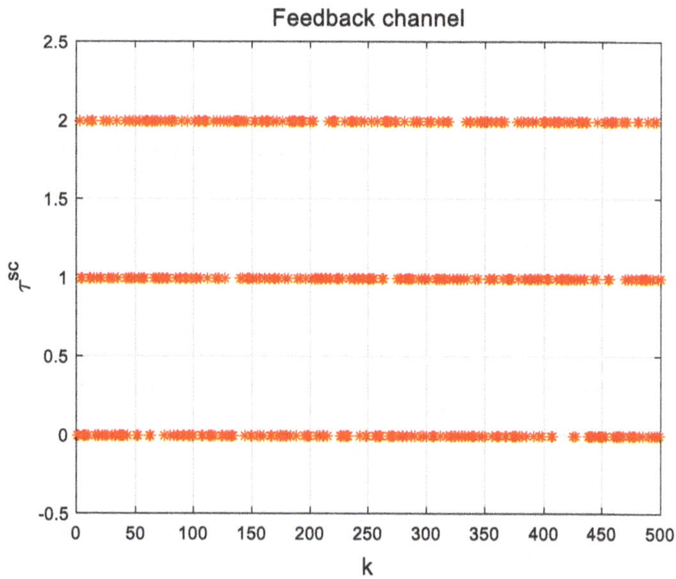

Fig. 3.4 Delay in feedback channel

Case 2: Random delay in the feedback channel and feedforward channel. In this case, for all $k \in \{1, 2, ..., 500\}$, $\overline{\tau}^{sc}(k) \in \{0, 1\}$,$\overline{\tau}^{ca}(k) \in \{0, 1, 2\}$.

Figures 3.4, 3.5, 3.6, and 3.7 show the simulation results of the feedback delay, feedforward delay, input, and output, respectively. A comparison of the compensated and uncompensated output curves in Fig. 3.7 indicates that the proposed method can effectively compensate for random delays in both channels.

Jitter is a serious problem in motor control, especially when the desired output is changed. As shown in Figs. 3.3, 3.4, 3.5, 3.6 and 3.7, the proposed method ensures good performance when the system is affected by latency and packet loss.

3.5 Conclusions

This chapter focused on the control of a DC-PMM in an NCS and the effect of network-induced delay and data packet loss on the control performance. The control structure of the DC-PMM was designed to address random delay and packet loss of the feedback channel and feedforward channel. A sliding mode predictive tracking control algorithm based on PPD was proposed. The algorithm only uses the measured input and output to compensate for the random delay and packet loss in the motor control. Two examples were presented to verify the performance of the proposed method, constant delay and random delay of the two channels. The simulation results showed that the proposed method compensated for the delay in the feedback channel

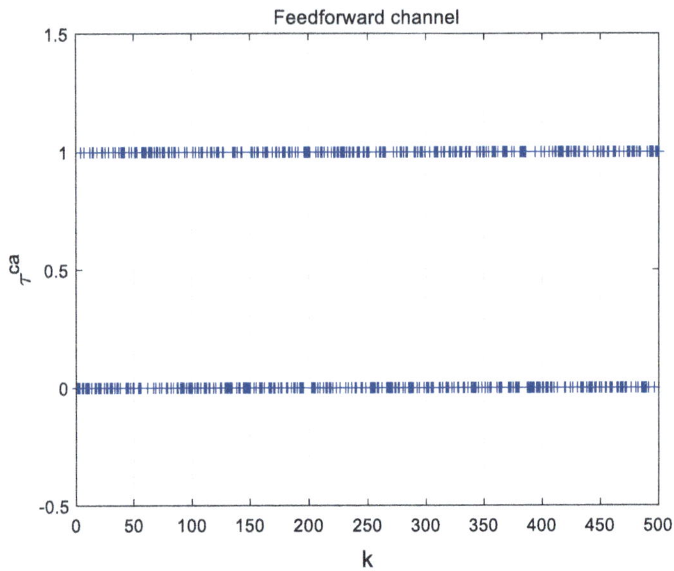

Fig. 3.5 Delay in feedforward channel

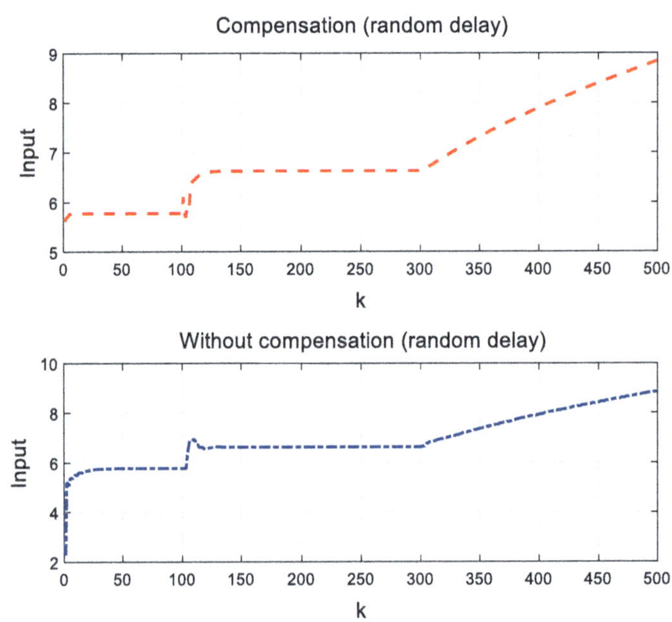

Fig. 3.6 Input of the plant with random delay

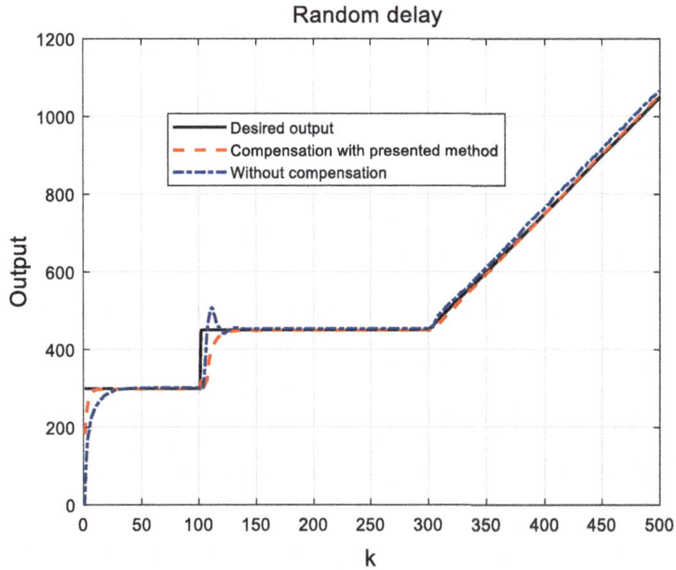

Fig. 3.7 Output of the plant with random delay

or the feedforward channel for constant or random delays. The output of the DC motor could track the desired trajectory rapidly, and the system remained stable.

References

1. M. Li, and Y. Chen, "Sliding mode predictive tracking control for dc permanent magnet motor in NCSs with random delay and packet dropouts," International Journal of Control, Automation and Systems, **15**(6), 2723-2732 (2017)
2. D. H. Zhai, and Y. Q. Xia, "Adaptive control for teleoperation system with varying time-delays and input saturation constraints," IEEE Transactions on Industrial Electronics, **63**(11), 6921–6929 (2016)
3. D. H. Zhai, and Y. Q. Xia, "Adaptive control of semi-autonomous teleoperation system with asymmetric time-varying delays and input uncertainties," IEEE Transactions on Cybernetics, **47**(11), 6921–6929 (2017)
4. Y. Q. Xia, Y. L. Gao, L. P. Yan, and M. Y. Fu, "Recent progress in networked control systems--A survey," International Journal of Automation and Computing, **12**(4), 343-367 (2015)
5. K. Liu, E. Fridmanb, and K. H. Johansson, "Dynamic quantization of uncertain linear networked control systems," *Automatica*, **59**, 248-255 (2015)
6. H. J. Gao, and T. W. Chen, "$H\infty$ estimation for uncertain systems with limited communication capacity," *IEEE Transactions on Automatic Control*, **52**(11), 2070-2084 (2007)
7. R. A. Gupta, and M. Y. Chow, "Networked control system: overview and research trends," IEEE Transactions on Industrial Electronics, **57**(7), 2527-2535 (2010)
8. Z. S. Hou, and X. H. Bu, "Model free adaptive control with data dropouts," Expert Systems with Applications, **38**(8), 10709-10717 (2011)

9. Z. H. Pang, G. P. Liu, D. H. Zhou, and D. H. Sun, "Data-based predictive control for networked non-linear systems with two-channel packet dropouts," IET Control Theory Application, **9**(7), 1154-1161 (2015)
10. H. B. Song, G. P. Liu, and L. Yu, "Networked predictive control of uncertain systems with multiple feedback channels," *IEEE Transactions on Industrial Electronics*, **60**(11), 5228-5238 (2013)
11. Y. Q. Xia, G. P. Liu, M. Fu, and D. Rees, "Predictive control of networked systems with random delay and data dropout," *IET Control Theory Application*, **3**(11), 1476–1486 (2009)
12. Y. M. Liu, and I. K. Fong, "Robust predictive tracking control of networked control systems with time-varying delays and data dropouts," *IET Control Theory Application*, **7**(5), 738-748 (2013)
13. G. P. Liu, Y. Q. Xia; J. Chen, D. Rees, and W. S. Hu, "Networked predictive control of systems with random network delays in both forward and feedback channels," *IEEE Transactions on Industrial Electronics*, **54**(3), 1282-1296 (2007)
14. J. H. Zhang, Y. Q. Xia, and P. Shi, "Design and stability analysis of networked predictive control systems," IEEE Transactions on Control Systems Technology, **21**(4), 1495-1501 (2013)
15. S. H. Li, and Z. G. Liu, "Adaptive speed control for permanent-magnet synchronous motor system with variations of load inertia," IEEE Transactions on Industrial Electronics, **56**(8), 3050-3059 (2009)
16. F. B. Carbajal, A. V. Gonzalez, J. C. Rosas-Caro, and A. Favela-Contreras, "An asymptotic differentiation approach of signals in velocity tracking control of DC motors," *Electric Power Systems Research*, **122**, 218-223 (2015)
17. R. S. Ortigoza, V. M. H. Guzm´an, M. Antonio-Cruz, and D. Munoz-Carrillo, "DC/DC buck power converter as a smooth starter for a dc motor based on a hierarchical control," *IEEE Transactions on Power Electronics*, **30**(2), 1076–1084 (2015)
18. J. P. Yu, Peng Shi, H. S. Yu, B. Chen, and C. Lin, "Approximation-based discrete-time adaptive position tracking control for interior permanent magnet synchronous motors," *IEEE Transactions on Cybernetics*, **45**(7), 1363–1371 (2015)
19. H. B. Li, M. Y. Chow, and Z. Q. Sun, "EDA-based speed control of a networked dc motor system with time delays and packet losses," *IEEE Transactions on Industrial Electronics*, **56**(5), 1727-1735 (2009)
20. X. Y. Zhu, H. Zhang, and Z. D. Fang, "Speed synchronization control for integrated automotive motor–transmission power train system with random delays," *Mechanical Systems and Signal Processing*, **64**(65), 46-57 (2015)
21. W. B. Gao, Y. F. Wang, and A. Homaifa, "Discrete-time variable structure control systems," *IEEE Transactions on Industrial Electronics*, **42**(2), 117–122 (1995)
22. S. C. Qu, X. H. Xia, and J. F. Zhang, "Dynamics of discrete time sliding-mode-control uncertain systems with a disturbance compensator," *IEEE Transactions on Industrial Electronics*, **61**(7), 3502–3510 (2014)

Chapter 4
High-Order Sliding-Mode Tracking Control of Networked Systems with Communication Constraints

This chapter investigates the fast tracking control of the network control system under communication constraints (network-induced delay, packet loss, packet disorder) and external interference (Li and Chen in IEEE Transactions on Industrial Electronics 64:4037–4047, 2017 [1]). A networked predictive control algorithm based on k-order adaptive discrete-time sliding mode control (k-ADSMC) is proposed. According to the error and error difference of the tracking output, an adaptive sliding mode parameter adjustment method is designed to ensure fast convergence of the ADSMC. k-ADSMC is proposed to improve the robustness of tracking control. It uses the estimated disturbance as a weighting factor to synthesize the k-th sliding mode values. Finally, the stability of the system is analyzed, and the effectiveness of the method is verified using a DC motor servo system as an example.

4.1 Introduction

At the end of the twentieth century, the networked control system (NCS) was developed by integrating a network and a control system. Compared with the traditional point-to-point control systems, the NCS has the advantages of resource sharing, low cost, easy expansion, simple installation and maintenance, and high reliability. Therefore, in recent years, NCSs have been used in the smart grid [2], for big data [3], in cloud control systems [4], and in remote operating systems [5] and other industrial fields. However, wired or wireless networks have caused many communication restrictions. Various methods have been proposed to deal with these two problems. A representative method is network predictive control (NPC), which actively compensates for the delay and packet loss in network communication [6–12]. The main objective of NPC is to predict the future behavior of the system based on the real-time information of delay and packet loss and take corresponding control measures.

At present, most studies of NCSs focused on stability analysis and state estimation [13, 14]. In contrast, tracking issues have received relatively little attention. In [13],

© The Author(s), under exclusive license to Springer Nature Singapore Pte Ltd. 2022
M. Li et al., *Tracking Control of Networked Systems via Sliding-Mode*,
https://doi.org/10.1007/978-981-16-6514-1_4

the robust H2 and H-infinity step tracking control methods were investigated, and a Markov chain was used to model the random time delay. However, only the network induction delay was considered, and the Markov transition matrix has to be known. In [14], Zhang et al. analyzed the step tracking control of a discrete-time nonlinear system. The Takagi–Sugeno fuzzy system was used to represent the nonlinear system, and the network-induced delay was considered in modeling the connection link. However, the author assumed that the lumped network-induced delay was a known set, and the probability of each element in the set was known. References [15-17] investigated H-infinity control of NCSs with network-induced delay and packet loss and designed a state feedback controller with a fixed gain to ensure good performance of the H-infinity output tracking. However, steady-state tracking errors occurred, and the system state was measured online, which may not be possible in practice. Pang et al. [18] studied the output tracking of the national statistical system. The output tracking error and incremental state were selected to form a new augmented system, and the output tracking problem was transformed into a stabilization problem of the augmented system. However, in practical applications, the incremental form of the system is difficult to obtain, especially for continuous-time systems with uncertain parameters and external disturbances.

Sliding mode control (SMC) is a non-linear and robust control strategy that guarantees perfect tracking performance even in systems with communication constraints and external interference. Discrete-time SMC (DSMC), an implementation of digital systems, is widely used in a variety of areas [19]. In [20], a nonlinear speed control algorithm for a permanent magnet synchronous motor (PMSM) servo system with SMC and disturbance compensation was developed. Reference [21] proposed a new DSMC method based on transient control. In [22], Veselí et al. proposed the discrete-time sliding mode (DSM) framework, a numerically controlled position system that used Euler velocity estimation. However, if the disturbance does not meet the matching criteria, its effects can only be partially offset; thus, the control design for systems with inconsistent disturbances remains a difficult problem. In [23], a second-order discrete-time terminal sliding mode control strategy (2-DTSMC) was designed. Unlike control based on state feedback, the reported 2-DTSMC scheme was developed based solely on output feedback. However, it was difficult to select the parameters for the terminal slide function. Since 2-DTSMC was not changed after the initial setting, the convergence speed was relatively slow.

Based on the above analysis, this paper proposes a new NPC algorithm called k-ADSMC for robust tracking control of NCSs with communication constraints and external interference. The method compensates for communication restrictions and external interference. ADSMC is designed for error-to-error errors in the tracking output, and the sliding mode parameters can be adaptively adjusted to ensure faster convergence. In addition, the k-ADSMC is designed to use the estimated interference as a weighting factor and integrate all k-th sliding mode values to improve tracking control robustness.

The rest of this chapter is organized as follows. The first part discusses robust tracking control issues of NCSs with communication constraints and external interference. The third section describes the k-ADSMC algorithm. The fourth section presents the proposed NPC method based on k-ADASMC. In Sect. 4.5, we describe a simulation experiment to demonstrate the effectiveness of the proposed algorithm. The conclusions are given in Sect. 4.6.

4.2 Problem Formulation

4.2.1 System Description

Consider a continuous-time linear system describes as follows; its structure is shown in Fig. 4.1:

$$\dot{\overline{x}}(t) = \overline{A}\overline{x}(t) + \overline{B}u(t) + \overline{D}f(t) \tag{4.1}$$

$$\overline{y}(t) = \overline{C}\overline{x}(t) \tag{4.2}$$

where $\overline{x}(t) \in R^n$, $u(t) \in R^m$, and $\overline{y}(t) \in R^q$ represent the system state, control input, and system output, respectively. $f(t) \in R^p$ represents the system parameters, control uncertainties, and external disturbances, and $f(t)$ can be regarded as a boundary; $\overline{A}$, $\overline{B}$, $\overline{C}$, and $\overline{D}$ are the constant matrices of the system with appropriate dimensions.

We assume that the control law u is established using a zero-order hold (ZOH) method, i.e., $u(t) = u(k)$ in the time interval $[kh, (k+1)h]$, where h is the sampling period. Therefore, the system can be discretized as [21]:

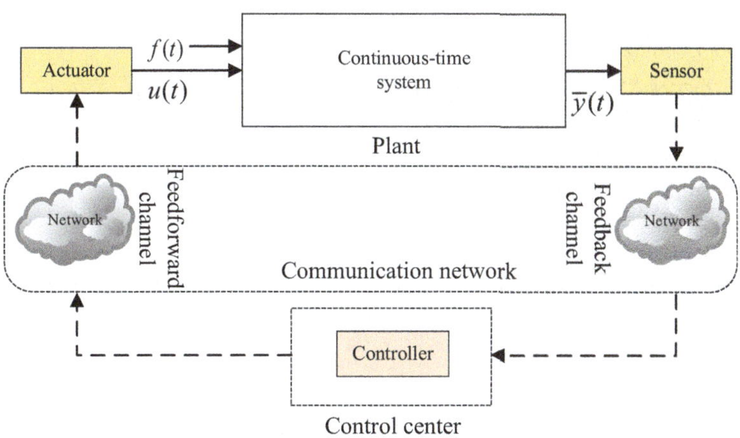

Fig. 4.1 Structure of the NCS

$$\overline{x}(k+1) = \overline{\Omega}\overline{x}(k) + \overline{\Psi}u(k) + \overline{F}(k) \tag{4.3}$$

$$\overline{y}(k) = \overline{\Gamma}\overline{x}(k) \tag{4.4}$$

where $\overline{\Omega} = e^{\overline{A}h}$, $\overline{\Psi} = \int_0^h e^{\overline{A}t}\overline{B}dt$, $\overline{\Gamma} = \overline{C}$, and $\overline{F}(k) = \int_0^h e^{\overline{A}t}\overline{D}f((k+1)h-t)dt$.

4.2.2 Disturbance Estimation

The disturbance term $\overline{F}(k)$ in Eq. (4.3) can be estimated by its one-step delayed value as follows:

$$\hat{\overline{F}}(k) = \overline{F}(k-1) = \overline{x}(k) - \overline{\Omega}\overline{x}(k-1) - \overline{\Psi}u(k-1) \tag{4.5}$$

Hence, according to the state-space models (9.3) and (9.4), the following is derived:

$$\overline{x}(k+1) = \overline{\Omega}\overline{x}(k) + \overline{\Psi}u(k) + \hat{\overline{F}}(k) - \tilde{\overline{F}}(k) \tag{4.6}$$

$$\overline{y}(k) = \overline{\Gamma}\overline{x}(k) \tag{4.7}$$

where $\tilde{\overline{F}}(k) = \hat{\overline{F}}(k) - \overline{F}(k)$ is the error of disturbance estimation. Based on Eqs. (4.3) and (4.5), $\tilde{\overline{F}}(k)$ can be written in the following form:

$$\begin{aligned}\tilde{\overline{F}}(k) &= \overline{F}(k-1) - \overline{F}(k)\\ &= \overline{\Omega}[\overline{x}(k) - \overline{x}(k-1)] + \overline{\Psi}[u(k) - u(k-1)] - [\overline{x}(k+1) - \overline{x}(k)]\end{aligned} \tag{4.8}$$

That is

$$\tilde{\overline{F}}(k) = \overline{\Omega}\Delta\overline{x}(k) + \overline{\Psi}\Delta u(k) - \Delta\overline{x}(k+1) \tag{4.9}$$

where $\Delta\overline{x}(k+1) = \overline{x}(k+1) - \overline{x}(k)$, $\Delta\overline{x}(k) = \overline{x}(k) - \overline{x}(k-1)$, and $\Delta u(k) = u(k) - u(k-1)$. It can be seen that Eq. (4.9) is the incremental form of Eq. (4.3).

4.3 Adaptive Sliding-Mode Control Algorithm

DSMC is a nonlinear and robust control strategy that guarantees perfect tracking performance even the system experiences communication delay and disturbance. The 2-DTSMC is designed first, and the K-order algorithm is designed second.

4.3.1 *Design of 2-order Sliding Function*

The tracking output error is defined as follows:

$$\overline{e}(k) = \overline{y}(k) - ref(k) \tag{4.10}$$

where $ref(k)$ denotes the desired output of the controlled system.

According to the output error defined in Eq. (4.10), the discrete-time sliding function is designed as follows:

$$\overline{S}(k) = \overline{S}(k-1) + \overline{\alpha}\overline{e}(k) + \overline{\beta} \times \mathrm{sgn}[-\overline{S}(k-1)] \times [\overline{e}(k-1)]^{\overline{\chi}} \tag{4.11}$$

where $\overline{\alpha}, \overline{\beta}$ are the weight parameters, $\overline{\alpha}, \overline{\beta} > 0$. sgn[·] represents the sign function,$0 < \overline{\chi} < 1$.

In Eq. (4.11), the parameter $\overline{\chi}$ produces a steep slope of the sliding surface near the origin of the phase plane. The direction of the sliding movement can be changed using the sign of the sliding mode function of the previous moment sgn[·]. This approach speeds up the process of reaching the sliding surface and increases the convergence.

In addition, the following 2nd-order sliding mode function that considers the influence of disturbance is designed:

$$\overline{\sigma}(k) = \overline{S}(k) + \overline{\omega}_k \overline{S}(k-1) \tag{4.12}$$

where $\overline{\omega}_k = 1/[1 + \overline{\gamma}\exp(-||\hat{\overline{F}}(k-1)||)]$, $0 < \overline{\gamma} < 1$, $0 \le \overline{\omega}_k \le 1$, and $||\cdot||$ denotes the 2-norm of the vector or matrix. The weight parameter $\overline{\omega}_k$ is decided by $||\hat{\overline{F}}(k-1)||$. As the value of $||\hat{\overline{F}}(k-1)||$ increases, the weight $\overline{\omega}_k$ increases, and the work of $\overline{S}(k-1)$ becomes more important. Especially, when $\overline{\omega}_k$ approaches to 0, the work of $\overline{S}(k-1)$ can be ignored, and the 2nd-order sliding mode is transformed into the first-order sliding mode.

4.3.2 Design of 2-order Sliding-Mode Controller

In this part, the equivalent control is used to design the sliding mode controller.

Let $\overline{\sigma}(k+1) = 0$, that is:

$$\overline{S}(k+1) + \overline{\omega}_{k+1}\overline{S}(k) = 0 \tag{4.13}$$

By considering Eq. (4.11), it can be deduced that:

$$\overline{S}(k) + \overline{\alpha}[\overline{y}(k+1) - ref(k+1)] + \overline{\beta} \times \text{sgn}[-\overline{S}(k)] \times [\overline{e}(k)]^{\overline{\chi}} = -\overline{\omega}_{k+1}\overline{S}(k) \tag{4.14}$$

By substituting the system model into Eq. (4.18), then:

$$\begin{aligned}&\overline{\alpha}\{\overline{\Gamma}[\overline{\Omega}\overline{x}(k) + \overline{\Psi}u(k) + \hat{\overline{F}}(k) - \tilde{\overline{F}}(k)] - ref(k+1)\} \\ &+ \overline{\beta} \times \text{sgn}[-\overline{S}(k)] \times [\overline{e}(k)]^{\overline{\chi}} = -(\overline{\omega}_{k+1} + 1)\overline{S}(k)\end{aligned} \tag{4.15}$$

When the error of the disturbance estimation $\tilde{\overline{F}}(k) = 0$, the equivalent control law $u_{eq}(k)$ is obtained:

$$\begin{aligned}u_{eq}(k) = {}& (\overline{\alpha}\overline{\Gamma}\overline{\Psi})^{-1}\{\overline{\alpha}ref(k+1) - \overline{\alpha}\overline{\Gamma}[\overline{\Omega}\overline{x}(k) + \hat{\overline{F}}(k)] \\ & - (\frac{1}{1 + \overline{\gamma}\exp(-||\hat{\overline{F}}(k)||)} + 1)\overline{S}(k) - \overline{\beta} \times \text{sgn}[-\overline{S}(k)] \times [\overline{e}(k)]^{\overline{\chi}}\}\end{aligned} \tag{4.16}$$

Robustness is ensured by adding discontinuities (the notation for the new slider function $\overline{\sigma}(k)$). By analogy to the continuous-time case, the integral of the discontinuous term approximated by the first-order transformation is applied to the system.

$$u_{dis}(k) = u_{dis}(k-1) - (\overline{\Gamma}\overline{\Psi})^{-1}\overline{\kappa}sign(\overline{\sigma}(k)) \tag{4.17}$$

where $sign[\cdot]$ is the signum function, and $\overline{\kappa} > 0$ is the switching control gain.

Therefore, the final control law that combines the equivalent control $u_{eq}(k)$ and the switching control $u_{dis}(k)$ is defined as follows:

$$u(k) = u_{eq}(k) + u_{dis}(k) \tag{4.18}$$

4.3.3 Design of K-order Adaptive Sliding Function

The tracking output error is defined as follows:

$$\overline{e}(k) = \overline{y}(k) - ref(k) \tag{4.19}$$

where $ref(k)$ denotes the desired output of controlled system.

According to the output error defined in Eq. (4.19), we design the adaptive discrete-time sliding function as follows:

$$\overline{S}(k) = \overline{S}(k-1) + g_1(\Delta\overline{e}(k-1))\overline{e}(k) + g_2(\Delta\overline{e}(k-1))(\overline{e}(k-1))^{\overline{\alpha}} \tag{4.20}$$

in which $g_1(\Delta\overline{e}(k-1)) = \frac{1}{1+\exp(-\overline{\beta}||\Delta\overline{e}(k-1)||)}$, $g_2(\Delta\overline{e}(k-1)) = \frac{1}{1+\overline{\gamma}||\Delta\overline{e}(k-1)||}$, $\Delta\overline{e}(k-1) = \overline{e}(k-1) - \overline{e}(k-2)$, $\overline{\beta}, \overline{\gamma} > 0$, and $0 < \overline{\alpha} < 1$.

Different from the traditional sliding mode function, a nonlinear term $(\overline{e}(k-1))^{\overline{\alpha}}$ and an adaptive function $g_1(\cdot)$ and $g_2(\cdot)$ are introduced. When the trajectory is unstable, i.e., $||\Delta\overline{e}(k-1)||$ increases, $g_1(\Delta\overline{e}(k-1))$ approaches 1, and $g_2(\Delta\overline{e}(k-1))$ approaches 0. Accordingly, in Eq. (4.11), the second term on the right mainly works. In contrast, when the trajectory is almost stable, i.e., $||\Delta\overline{e}(k-1)||$ decreases, $g_1(\Delta\overline{e}(k-1))$ approaches 0, and $g_2(\Delta\overline{e}(k-1))$ approaches 1; thus, the third term on the right side of Eq. (4.11) mainly works. Thus, the proposed sliding function not only to ensures fast convergence but also reduces jitter.

Furthermore, a k-order sliding mode function is designed as follows to suppress the effect of disturbance:

$$\begin{aligned}\overline{\sigma}(k) &= \overline{S}(k) + \omega_{k-1}\overline{S}(k-1) + \omega_{k-2}\overline{S}(k-2) + \ldots + \omega_1\overline{S}(1) \\ &= \overline{S}(k) + \frac{||\hat{\overline{F}}(k-1)||}{\sum_{i=1}^{k-1}||\hat{\overline{F}}(i)||}\overline{S}(k-1) + \frac{||\hat{\overline{F}}(k-2)||}{\sum_{i=1}^{k-1}||\hat{\overline{F}}(i)||}\overline{S}(k-2) \\ &+ \ldots + \frac{||\hat{\overline{F}}(1)||}{\sum_{i=1}^{k-1}||\hat{\overline{F}}(i)||}\overline{S}(1) = \overline{S}(k) + \sum_{j=1}^{k-1}\frac{||\hat{\overline{F}}(j)||}{\sum_{i=1}^{k-1}||\hat{\overline{F}}(i)||}\overline{S}(j)\end{aligned} \tag{4.21a}$$

where $||\cdot||$ denotes the 2-norm of the vector or matrix. Thus, $\sum_{j=1}^{k-1}\omega_j = 1$ and $0 \le \omega_j \le 1$.

In Eq. (4.21a), the weight ω_j is determined by the proportion of $||\hat{\overline{F}}(j)||$. If the proportion of $||\hat{\overline{F}}(j)||$ is high, the weight ω_j is large and vice versa. The k-order sliding mode function $\overline{\sigma}(k)$ uses the newest k sliding mode function $\{\overline{S}(k), \overline{S}(k-1), \ldots, \overline{S}(1)\}$, which uses ω_j as a weight factor to determine the contribution of $\overline{S}(j)$. For example, if ω_j is large, it indicates that the proportion of $||\hat{\overline{F}}(j)||$ is high, and the work of $\overline{S}(j)$ is important. In contrast, if ω_j is small, the work of $\overline{S}(j)$ is negligible. When ω_j approaches 0, the work of $\overline{S}(j)$ can be ignored. The k-order sliding mode function selects the estimated disturbance as feedback information to suppress the effect of the external disturbance.

Let $\breve{\overline{F}}(k-1)=\sum_{j=1}^{k-1}[||\hat{\overline{F}}(j)||/(\sum_{i=1}^{k-1}||\hat{\overline{F}}(i)||)]\overline{S}(j)$, then Eq. (4.21a) can be written as follows:

$$\overline{\sigma}(k)=\overline{S}(k)+\breve{\overline{F}}(k-1) \tag{4.21b}$$

4.3.4 Design of Sliding-Mode Controller

The equivalent control in [24] is used to design the sliding mode controller. Let $\overline{\sigma}(k+1)=0$, i.e.:

$$\overline{S}(k+1)+\breve{\overline{F}}(k)=0 \tag{4.22}$$

So that

$$\overline{S}(k+1)=-\breve{\overline{F}}(k) \tag{4.23}$$

In combination with Eq. (4.11), it can be deduced that:

$$\breve{\overline{F}}(k)=-\overline{S}(k)-g_1(\Delta\overline{e}(k))[\overline{y}(k+1)-ref(k+1)]-g_2(\Delta\overline{e}(k))(\overline{e}(k))^{\overline{\alpha}} \tag{4.24}$$

By substituting Eqs. (4.6) and (4.7) into Eq. (4.15), we obtain:

$$\begin{aligned}\breve{\overline{F}}(k)=&-\overline{S}(k)-g_1(\Delta\overline{e}(k))[\overline{\Gamma}(\overline{\Omega}\overline{x}(k)+\overline{\Psi}u(k)\\&+\hat{\overline{F}}(k)-\tilde{\overline{F}}(k))-ref(k+1)]-g_2(\Delta\overline{e}(k))(\overline{e}(k))^{\overline{\alpha}}\end{aligned} \tag{4.25}$$

When the error of the disturbance estimation $\tilde{\overline{F}}(k)=0$, we solve Eq. (4.25) to obtain the equivalent control $u_{eq}(k)$:

$$\begin{aligned}&u_{eq}(k)\\&=[g_1(\Delta\overline{e}(k))\overline{\Gamma\Psi}]^{-1}[g_1(\Delta\overline{e}(k))ref(k+1)-\overline{S}(k)-\breve{\overline{F}}(k)\\&-g_2(\Delta\overline{e}(k))(\overline{e}(k))^{\overline{\alpha}}-g_1(\Delta\overline{e}(k))\overline{\Gamma\Omega}\overline{x}(k)-g_1(\Delta\overline{e}(k))\overline{\Gamma}\hat{\overline{F}}(k)]\end{aligned} \tag{4.26a}$$

That is:

$$u_{eq}(k) = [1 + \exp(-\overline{\beta}||\Delta\overline{e}(k)||)](\overline{\Gamma\Psi})^{-1}\{\frac{1}{1 + \exp(-\overline{\beta}||\Delta\overline{e}(k)||)}$$
$$\times [ref(k+1) - \overline{\Gamma\Omega}\overline{x}(k) - \overline{\Gamma}\hat{\overline{F}}(k)] - \frac{1}{1 + \overline{\gamma}||\Delta\overline{e}(k)||}(\overline{e}(k))^{\overline{\alpha}}$$
$$- \overline{S}(k) - \sum_{j=1}^{k}\frac{||\hat{\overline{F}}(j)||}{\sum_{i=1}^{k}||\hat{\overline{F}}(i)||}\overline{S}(j)\} \tag{4.26b}$$

The robustness is ensured by the addition of a discontinuous term (sign of the new sliding function $\overline{\sigma}(k)$). Similar to the continuous-time case, the integral of the discontinuous term approximated by the first-order transformation is applied to the systems described in Eqs. (4.3) and (4.4) [25]:

$$u_{dis}(k) = u_{dis}(k-1) - (\overline{\Gamma\Psi})^{-1}\kappa sign(\overline{\sigma}(k)) \tag{4.27}$$

where $sign[\cdot]$ is the signum function, and $\kappa > 0$ is the switching control gain.

Therefore, the equivalent control $u_{eq}(k)$ is combined with the switching control $u_{dis}(k)$ to create the overall control law:

$$u(k) = u_{eq}(k) + u_{dis}(k) \tag{4.28}$$

4.3.5 Analysis of Stability

In a robust tracking control system, such as the DC motor servo system, the state $\overline{x}(k)$ usually denotes the current and angular velocity, the input $u(k)$ is the voltage. Both are bounded, and the matrix A in the system (4.1) has eigenvalues with negative real parts. In Eq. (4.9), $\overline{\Omega}$ and $\overline{\Psi}$ are constant matrices; hence, when the input $u(k)$ and state $\overline{x}(k)$ are limited, it is reasonable to assume that $\tilde{\overline{F}}(k)$ is bounded.

Assumption 1 The error of disturbance estimation $\tilde{\overline{F}}(k)$ is bounded, and $|\tilde{\overline{F}}(k)| \leq \eta$, where$\eta > 0$.

Another assumption is introduced to facilitate the stability analysis. Let $\rho = \min_k(g_1(\Delta\overline{e}(k)))$, then $0 < \rho < 1$.

Assumption 2 The following condition is satisfied:

$$|g_1(\Delta\overline{e}(k-1))\overline{\Gamma}\tilde{\overline{F}}(k-1) - g_1(\Delta\overline{e}(k))\Gamma\tilde{\overline{F}}(k) + \Delta g_1(\Delta\overline{e}(k))\overline{\Gamma\Psi}u_{dis}(k)| < \mu < \rho\kappa$$

where $\mu > 0$, and $\Delta g_1(\Delta\overline{e}(k)) = g_1(\Delta\overline{e}(k)) - g_1(\Delta\overline{e}(k-1))$.

Remark 1 Since $0 < g_1(\Delta\bar{e}(k-1)), g_1(\Delta\bar{e}(k)) < 1$, hence $\Delta g_1(\Delta\bar{e}(k))$ is bounded as well. Based on Assumption 1, $\tilde{\bar{F}}(k-1)$ and $\tilde{\bar{F}}(k)$ are bounded; thus, Assumption 2 is reasonable.

Theorem 1 *For the system described by Eqs.* (4.3) *and* (4.4) *with the sliding function* (4.21b) *and Assumptions* 1 *and* 2, *a quasi-sliding mode with a bandwidth of* $|\bar{\sigma}(k)| < \mu + \rho\kappa$ *will occur within finite steps when the control law* (4.28) *is used.*

Proof According to Eqs. (4.19) and (4.28), the one-step forward tracking output error is derived as follows:

$$
\begin{aligned}
&\bar{e}(k+1) \\
&= \bar{y}(k+1) - ref(k+1) \\
&= \overline{\Gamma\Omega}\bar{x}(k) + \overline{\Gamma\Psi}u(k) + \overline{\Gamma F}(k) - ref(k+1) \\
&= \overline{\Gamma\Omega}\bar{x}(k) + \overline{\Gamma\Psi}u(k) + \overline{\Gamma F}(k) - ref(k+1) \\
&= \overline{\Gamma\Omega}\bar{x}(k) + \overline{\Gamma\Psi}(u_{eq}(k) + u_{dis}(k)) + \overline{\Gamma F}(k) - ref(k+1) \\
&= \overline{\Gamma\Omega}\bar{x}(k) + \overline{\Gamma\Psi}\{(\mathrm{g}_1(\Delta\bar{\mathrm{e}}(\mathrm{k}))\overline{\Gamma\Psi})^{-1}[\mathrm{g}_1(\Delta\bar{\mathrm{e}}(\mathrm{k}))\mathrm{ref}(\mathrm{k}+1) \\
&\quad - \bar{S}(k) - \breve{\bar{F}}(k) - g_2(\Delta\bar{e}(k))(\bar{e}(k))^{\bar{\alpha}} - g_1(\Delta\bar{e}(k))\overline{\Gamma\Omega}x(k) \\
&\quad - g_1(\Delta\bar{e}(k))\overline{\Gamma\hat{F}}(k)]\} + \overline{\Gamma\Psi}u_{dis}(k) + \overline{\Gamma F}(k) - ref(k+1) \\
&= \overline{\Gamma\Omega}x(k) + ref(k+1) - (g_1(\Delta\bar{e}(k)))^{-1}\bar{S}(k) \\
&\quad - (g_1(\Delta\bar{e}(k)))^{-1}\breve{\bar{F}}(k) - (g_1(\Delta\bar{e}(k)))^{-1}g_2(\Delta\bar{e}(k))(\bar{e}(k))^{\bar{\alpha}} \\
&\quad - \overline{\Gamma\Omega}x(k) - \overline{\Gamma\hat{F}}(k) + \overline{\Gamma\Psi}u_{dis}(k) + \overline{\Gamma F}(k) - ref(k+1) \\
&= \bar{\Gamma}[\bar{F}(k) - \hat{\bar{F}}(k)] - (g_1(\Delta\bar{e}(k)))^{-1}[\bar{S}(k) + \breve{\bar{F}}(k) \\
&\quad + g_2(\Delta\bar{e}(k))(\bar{e}(k))^{\bar{\alpha}}] + \overline{\Gamma\Psi}u_{dis}(k)
\end{aligned}
\tag{4.29}
$$

Considering the k-order sliding function in Eq. (4.21b), it can be deduced that:

$$
\begin{aligned}
&\bar{\sigma}(k+1) \\
&= \bar{S}(k+1) + \breve{\bar{F}}(k) \\
&= \bar{S}(k) + g_1(\Delta\bar{e}(k))\bar{e}(k+1) + g_2(\Delta\bar{e}(k))(\bar{e}(k))^{\bar{\alpha}} + \breve{\bar{F}}(k) \\
&= \bar{S}(k) + g_1(\Delta\bar{e}(k))\{\bar{\Gamma}[\bar{F}(k) - \hat{\bar{F}}(k)] - (g_1(\Delta\bar{e}(k)))^{-1}[\bar{S}(k) + \breve{\bar{F}}(k) \\
&\quad + g_2(\Delta\bar{e}(k))(\bar{e}(k))^{\bar{\alpha}}] + \overline{\Gamma\Psi}u_{dis}(k)\} + g_2(\Delta\bar{e}(k))(\bar{e}(k))^{\bar{\alpha}} + \breve{\bar{F}}(k) \\
&= g_1(\Delta\bar{e}(k))\bar{\Gamma}[\bar{F}(k) - \hat{\bar{F}}(k)] + g_1(\Delta\bar{e}(k))\overline{\Gamma\Psi}u_{dis}(k) \\
&= g_1(\Delta\bar{e}(k))\overline{\Gamma\Psi}u_{dis}(k) - g_1(\Delta\bar{e}(k))\overline{\Gamma\tilde{F}}(k)
\end{aligned}
\tag{4.30}
$$

Then, it is deduced from the above formulas that:

$$\overline{\sigma}(k) = g_1(\Delta\overline{e}(k-1))\overline{\Gamma\Psi}u_{\text{dis}}(k-1) - g_1(\Delta\overline{e}(k-1))\overline{\Gamma\tilde{F}}(k-1) \tag{4.31}$$

According to Assumption 1, $\overline{\sigma}(k)$ is bounded. Thus, the difference between Eqs. (4.30) and (4.31) can be computed as follows:

$$\begin{aligned}
&\overline{\sigma}(k+1) - \overline{\sigma}(k) \\
&= g_1(\Delta\overline{e}(k))\overline{\Gamma\Psi}u_{dis}(k) - g_1(\Delta\overline{e}(k-1))\overline{\Gamma\Psi}u_{dis}(k-1) \\
&\quad + g_1(\Delta\overline{e}(k-1))\overline{\Gamma\tilde{F}}(k-1) - g_1(\Delta\overline{e}(k))\overline{\Gamma\tilde{F}}(k) \\
&= g_1(\Delta\overline{e}(k-1))\overline{\Gamma\Psi}[\mathrm{u_{dis}(k) - u_{dis}(k-1)} + \mathrm{g_1(\Delta\overline{e}(k-1))} \\
&\quad \times \overline{\Gamma\tilde{F}}(k-1) - g_1(\Delta\overline{e}(k))\Gamma\tilde{F}(k) + \Delta g_1(\Delta\overline{e}(k))\overline{\Gamma\Psi}u_{dis}(k) \\
&= g_1(\Delta\overline{e}(k-1))\overline{\Gamma\tilde{F}}(k-1) - g_1(\Delta\overline{e}(k))\overline{\Gamma\tilde{F}}(k) \\
&\quad + \Delta g_1(\Delta\overline{e}(k))\overline{\Gamma\Psi}u_{dis}(k) - g_1(\Delta\overline{e}(k-1))\kappa sign(\overline{\sigma}(k))
\end{aligned} \tag{4.32}$$

Additionally, considering Assumption 2, it is deduced from Eq. (4.32) that:

$$\begin{aligned}
&-\mu - g_1(\Delta\overline{e}(k-1))\kappa sign(\overline{\sigma}(k)) < \overline{\sigma}(k+1) \\
&-\overline{\sigma}(k) < \mu - g_1(\Delta\overline{e}(k-1))\kappa sign(\overline{\sigma}(k))
\end{aligned} \tag{4.33}$$

(i) If $\overline{\sigma}(k)$ is a positive sequence, i.e., $\overline{\sigma}(k) > 0$ for $\forall k$, then it can be deduced from Eq. (4.33) that:

$$\overline{\sigma}(k+1) - \overline{\sigma}(k) < \mu - g_1(\Delta\overline{e}(k-1))\kappa < \mu - \rho\kappa < 0 \tag{4.34}$$

Inequality (4.34) indicates $\overline{\sigma}(k+1) - \overline{\sigma}(k) < 0 \Rightarrow \overline{\sigma}(k+1) < \overline{\sigma}(k)$, since $\overline{\sigma}(k) > 0$; hence $|\overline{\sigma}(k+1)| < |\overline{\sigma}(k)|$.

(ii) Otherwise, if $\overline{\sigma}(k) < 0$ for $\forall k$:

$$0 < \rho\kappa - \mu < \overline{\sigma}(k+1) - \overline{\sigma}(k) \tag{4.35}$$

Inequality (4.35) indicates $0 < \overline{\sigma}(k+1) - \overline{\sigma}(k) \Rightarrow \overline{\sigma}(k) < \overline{\sigma}(k+1)$, since $\sigma(k) < 0$; therefore, $|\overline{\sigma}(k+1)| < |\overline{\sigma}(k)|$.

According to (i) and (ii), we obtain:

$$|\overline{\sigma}(k+1)| < |\overline{\sigma}(k)| \tag{4.36}$$

Inequality (4.36) indicates that the sliding function $\overline{\sigma}(k)$ will converge.

In the following, the sliding function $\overline{\sigma}(k)$ entering the quasi-sliding domain is defined as:

$$\overline{\Xi} = \{\overline{\sigma}(k) : |\overline{\sigma}(k)| < \mu + \rho\kappa\} \tag{4.37}$$

in a finite number of steps.

If we assume that $\overline{\sigma}(0) > \mu + \rho\kappa$, it can be derived from Eq. (4.36) that:

$$\overline{\sigma}(k) - \overline{\sigma}(k+1) > \rho\kappa - \mu > 0 \tag{4.38}$$

We use $\overline{\zeta}$, because of $\overline{\zeta} > \rho\kappa - \mu$. Therefore,

$$\overline{M} = \frac{\overline{\sigma}_0 - (\mu + \rho\kappa)}{\overline{\zeta}} < \frac{\overline{\sigma}_0 - (\mu + \rho\kappa)}{\rho\kappa - \mu} \tag{4.39}$$

$\overline{M}$ is a finite number; additionally, when the initial value $\overline{\sigma}(0) < -\mu - \rho\kappa$, a similar result can be obtained.

Once $\overline{\sigma}(k)$ has entered the quasi-sliding domain, it will remain in that domain. For $\forall k \in N^+$, we assume that $\overline{\sigma}(k)$ is a positive value in the quasi-sliding domain, i.e.,

$$0 \leq \overline{\sigma}(k) < \mu + \rho\kappa \tag{4.40}$$

Then it can be derived from Eq. (4.32) that:

$$\begin{aligned}\overline{\sigma}(k+1) = \overline{\sigma}(k) + [&g_1(\Delta\overline{e}(k-1))\overline{\Gamma\tilde{F}}(k-1) - g_1(\Delta\overline{e}(k))\overline{\Gamma\tilde{F}}(k) \\ &+ \Delta g_1(\Delta\overline{e}(k))\overline{\Gamma\Psi u_{dis}}(k)] - g_1(\Delta\overline{e}(k-1))\kappa\end{aligned} \tag{4.41}$$

Based on Assumption 2, the following relationship is derived:

$$\overline{\sigma}(k) - \mu - g_1(\Delta\overline{e}(k-1))\kappa < \overline{\sigma}(k+1) < \overline{\sigma}(k) + \rho\kappa - g_1(\Delta\overline{e}(k-1))\kappa \tag{4.42}$$

Based on Eq. (4.40), it is deduced that:

$$-\mu - g_1(\Delta\overline{e}(k-1))\kappa < \overline{\sigma}(k+1) < \mu + 2\rho\kappa - g_1(\Delta\overline{e}(k-1))\kappa \tag{4.43}$$

Since $g_1(\Delta\overline{e}(k-1)) > \rho > 0$, hence

$$-\mu - \rho\kappa < \overline{\sigma}(k+1) < \mu + \rho\kappa \tag{4.44}$$

Inequality (4.44) indicates that $\overline{\sigma}(k+1) \in \overline{\Xi}$. Similarly, if $\overline{\sigma}(k)$ is a negative value in the quasi-sliding domain, i.e., $-\mu - \rho\kappa < \overline{\sigma}(k) < 0$, the same relationship (4.44) is generated. This completes the proof.

4.4 Robust Tracking Control of Networked Systems Based on High-Order Sliding-Mode

The structure of the NCS with random delay and packet loss is shown in Fig. 4.2. It includes three parts: the data buffer and network compensator on the plant side and the prediction generator on the controller side. A data buffer that is simple to calculate is used to pack the data. The prediction generator is used to generate a series of control inputs based on k-ADSMC. The network compensator is used to select the appropriate control input for the plant.

In Fig. 4.2, the plant is a discrete-time system with communication constraints and external disturbance, which is described in Eqs. (4.3) and (4.4). The delays in the feedback channel and feed-forward channel are $\overline{\tau}^{sc}(k)$, and $\overline{\tau}^{ca}(k)$, respectively, and $\overline{\tau}^{c}(k)$ is the calculation delay of the controller. In Fig. 4.2, $ref(t)$ is the reference signal. When sending data packets from the sensor and controller, the data packets are time-stamped. If a data packet does not reach its destination within a certain transmission time, the data will be lost according to commonly used network protocols. It is assumed that only a limited number of consecutive data packet losses can be tolerated to avoid the NCS becoming an open loop. Usually, packet loss is determined by opening the switches $\overline{S}_1$ and $\overline{S}_2$. In this paper, the network-induced delay, packet disorder, and packet dropouts are treated as a round-trip time (RTT) delay $\overline{\tau}(k)$. The system satisfies the following assumptions.

Assumption 3 The random RTT delay $\overline{\tau}(k)$ is bounded, i.e., $0 \leq \overline{\tau}(k) \leq \widehat{\overline{\tau}}$.

Assumption 4 The sensor is time-driven, and the controller and actuator are driven by events. Each node is clock-synchronized.

Remark 2 The RTT delay $\overline{\tau}(k)$ represents the total time delay of data packets successfully arriving at the actuator in sequence. The total time delay in the actuator can

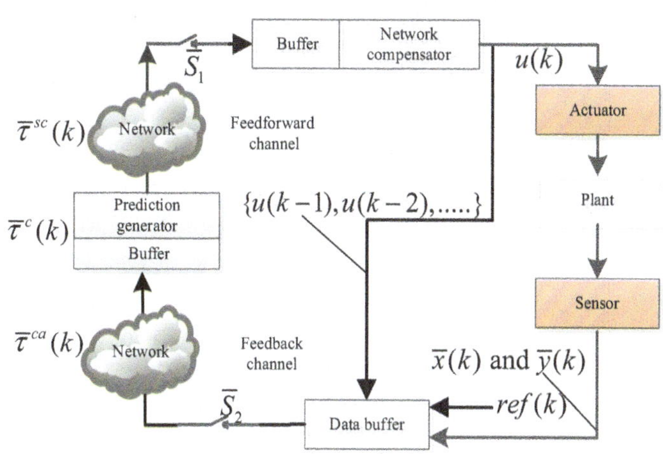

Fig. 4.2 Structure of NCS with random delay and packet dropouts

be obtained by subtracting the time stamp of the latest control data packet from the current time of the actuator at each sampling instant. As a result, it includes the network-induced delays, packet confusion, and packet loss in the two communication channels.

4.4.1 Data Buffer

At each sampling instant, the data buffer at the sensor side sends the following feedback packet together with the time stamp to the controller:

$$\bar{D}_{k_s} = [\bar{e}(k_s)^T, \bar{R}(k_s)^T, \bar{x}(k_s)^T, u(k_s)^T]^T \tag{4.45}$$

where $\overline{e}(k_s) = \overline{y}(k_s) - ref(k_s)$, $\overline{R}(k_s) = [ref(k_s)^T, ref(k_s+1)^T, ..., ref(k_s+\widehat{\overline{\tau}})^T]^T$, and k_s is the time stamp. It is assumed that $\overline{R}(k_s)$ is known at time k_s. It is known from Eq. (4.45) that the state $\overline{x}(k)$ and output $\overline{y}(k)$ of the system are included in the feedback information.

4.4.2 Prediction Generator

At the controller side, the buffer is used to record the latest control prediction sequence $\overline{D}_{k_s}$. The following state observer is designed to obtain the state vector of the plant at the controller side:

$$\hat{\overline{x}}(k_s+1|k_s) = \overline{\Omega}\hat{\overline{x}}(k_s|k_s-1) + \overline{\Psi}u(k_s|k_s-1) + L(\overline{y}(k_s) - \overline{\Gamma}\hat{\overline{x}}(k_s|k_s-1)) \tag{4.46}$$

$$\hat{\overline{y}}(k_s+1|k_s) = \overline{\Gamma}\hat{\overline{x}}(k_s+1|k_s) \tag{4.47}$$

where L is the observer gain matrix to be determined, $\hat{\overline{x}}(k_s+1|k_s)$ is the one-step state prediction vector, and $\hat{\overline{y}}(k_s+1|k_s)$ is the output vector of the observer.

The estimated value of the output tracking error is:

$$\hat{\overline{e}}(k_s+1|k_s) = ref(k_s+1) - \hat{\overline{y}}(k_s+1|k_s) \tag{4.48}$$

$$\Delta\hat{\overline{e}}(k_s+1|k_s) = \hat{\overline{e}}(k_s+1|k_s) - \hat{\overline{e}}(k_s|k_s) \tag{4.49}$$

Equations (4.26a), (4.26b), (4.27), and (4.28) indicate that the following one-step predictive value of control input can be derived:

$$u(k_s+1|k_s) = [1+\exp(-\overline{\beta}||\Delta\hat{\overline{e}}(k_s+1|k_s)||)] \times (\overline{\Gamma\Psi})^{-1}$$
$$\{\frac{1}{1+\exp(-\overline{\beta}||\Delta\hat{\overline{e}}(k_s+1|k_s)||)}[ref(k_s+1) - \overline{\Gamma\Omega}\hat{\overline{x}}(k_s+1)$$
$$-\overline{\Gamma}\hat{\overline{F}}(k_s+1)] - \frac{1}{1+\gamma||\Delta\hat{\overline{e}}(k_s+1|k_s)||} \times (\hat{\overline{e}}(k_s+1|k_s))^{\overline{\alpha}}$$
$$-\overline{S}(k_s+1) - \sum_{j=1}^{k_s+1}\frac{||\hat{\overline{F}}(j)||}{\sum\limits_{i=1}^{k_s+1}||\hat{\overline{F}}(i)||}\overline{S}(j)\} - (\overline{\Gamma\Psi})^{-1}\kappa sign(\overline{\sigma}(k_s+1)) \tag{4.50}$$

Based on the above results, the predictions of the state, output, tracking error, and control input up to time $k_s+\widehat{\overline{\tau}}$ are derived as follows:

$$\hat{\overline{x}}(k_s+i|k_s) = \overline{\Omega}\hat{\overline{x}}(k_s+i-1|k_s) + \overline{\Psi}u(k_s+i-1|k_s) \tag{4.51}$$

$$\hat{\overline{y}}(k_s+i|k_s) = \overline{\Gamma}\hat{\overline{x}}(k_s+i|k_s) \tag{4.52}$$

$$\hat{\overline{e}}(k_s+i|k_s) = ref(k_s+i) - \hat{\overline{y}}(k_s+i|k_s) \tag{4.53}$$

$$\Delta\hat{\overline{e}}(k_s+i|k_s) = \hat{\overline{e}}(k_s+i|k_s) - \hat{\overline{e}}(k_s+i-1|k_s) \tag{4.54}$$

$$u(k_s+i|k_s) = [1+\exp(-\beta||\Delta\hat{\overline{e}}(k_s+i|k_s)||)] \times (\overline{\Gamma\Psi})^{-1}$$
$$\times\{\frac{1}{1+\exp(-\overline{\beta}||\Delta\hat{\overline{e}}(k_s+i|k_s)||)}[ref(k_s+i) - \overline{\Gamma\Omega}\hat{\overline{x}}(k_s+i)$$
$$-\overline{\Gamma}\hat{\overline{F}}(k_s+i)] - \frac{1}{1+\gamma||\Delta\hat{\overline{e}}(k_s+i|k_s)||} \times (\hat{\overline{e}}(k_s+i|k_s))^{\overline{\alpha}}$$
$$-\overline{S}(k_s+i) - \sum_{j=1}^{k_s+i}\frac{||\hat{\overline{F}}(j)||}{\sum\limits_{l=1}^{k_s+i}||\hat{\overline{F}}(l)||}\overline{S}(j)\} - (\overline{\Gamma\Psi})^{-1}\kappa sign(\overline{\sigma}(k_s+i)) \tag{4.55}$$

where $i = 2, ..., \widehat{\overline{\tau}}$. According to Eqs. (4.46) to (4.55), the following control input sequence is generated:

$$\overline{U}_{k_s} = [u(k_s|k_s)^T, u(k_s+1|k_s)^T, \ldots, u(k_s+\widehat{\overline{\tau}}|k_s)^T]^T \tag{4.56}$$

Then $\overline{U}_{k_s}$ is packed with the time stamp k_s and transmitted to the actuator. Additionally, the controller performs continuous updates, as shown in Fig. 4.3 . When the true value arrives at the controller, the predicted value is replaced by the true value. Then, the new datum is used to predict the next time value.

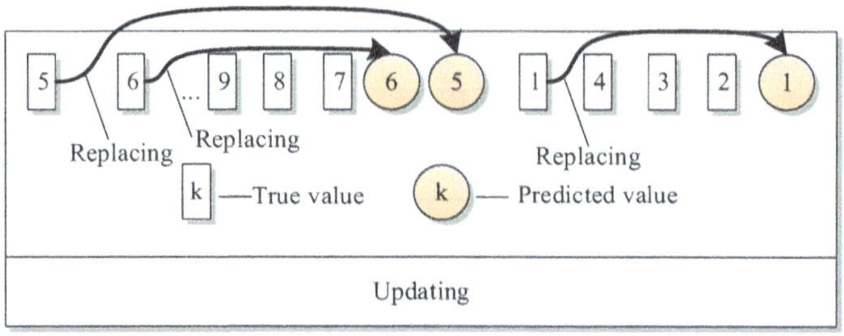

Fig. 4.3 The updating process of packet

4.4.3 Network Compensator

A network compensator is added to the actuator. Due to random delays in the forward channel, packet confusion, and packet loss, it is very likely that zero, one, or more control packets arrive at the actuator within a sampling interval. Therefore, the buffer is used to store the latest control prediction sequence $\overline{U}_{k_s}$, and it is assumed that the corresponding time stamp $\tilde{\bar{k}}_s$ is available in the actuator. The buffer performs the following update process at each sampling moment:

$$\tilde{\bar{k}}_s = \begin{cases} k_s, & \text{if}(\kappa_s \geq \tilde{\bar{\text{k}}}_\text{s}) \\ \tilde{\bar{k}}_s, & if(\kappa_s \leq \tilde{\bar{\text{k}}}_\text{s})|(\text{nopacketsarrive}) \end{cases} \tag{4.57}$$

$$\bar{U}_{\tilde{\bar{k}}_s} = \begin{cases} \bar{U}_{k_s}, & \text{if}(\kappa_s \geq \tilde{\bar{\text{k}}}) \\ \bar{U}_{\tilde{\bar{k}}_s}, & if(\kappa_s \leq \tilde{\bar{\text{k}}})|\ (\text{no packets arrive}) \end{cases} \tag{4.58}$$

where $\overline{U}_{\tilde{\bar{k}}_s}$ can be expressed as:

$$\overline{U}_{\tilde{\bar{k}}_s} = [u(\tilde{\bar{k}}_s|\tilde{\bar{k}}_s)^T, u(\tilde{\bar{k}}_s + 1|\tilde{\bar{k}}_s)^T, ..., u(\tilde{\bar{k}}_s + \hat{\bar{\tau}}|\tilde{\bar{k}}_s)^T]^T \tag{4.59}$$

The real-time RTT delay $\bar{\tau}(k)$ can be calculated in the actuator at the instant k as follows:

$$\bar{\tau}(k) = k - \tilde{\bar{k}}_s \tag{4.60}$$

where $\bar{\tau}(k) \leq \hat{\bar{\tau}}$ is a random nonnegative integer. The network compensator transmits the following control signal to the plant at time k to compensate for the RTT delay $\bar{\tau}(k)$:

$$u(k) = u(\tilde{\bar{k}}_s + \bar{\tau}(k)|\tilde{\bar{k}}_s) \tag{4.61}$$

4.4.4 Stability Analysis

Due to the random RTT delay $\overline{\tau}(k)$ in the NCS, according to the network compensator, the input of the plant at the instant k is $u(k) = u(\tilde{\overline{k}}_s + \overline{\tau}(k)|\tilde{\overline{k}}_s)$. By substituting Eq. (4.61) into Eq. (4.3), the following closed-loop NCS can be obtained:

$$\begin{aligned}&\overline{x}(k+1)\\&=\overline{\Omega}\overline{x}(k)+\overline{\Psi}u(\tilde{\overline{k}}_s+\overline{\tau}(k)|\tilde{\overline{k}}_s)+\overline{F}(k)\\&=\overline{\Omega}\overline{x}(k)+\overline{\Psi}(u(k)+\overline{F'}(k))\end{aligned}\tag{4.62}$$

Let $\overline{X}(k) = [\overline{x}(k)\,\overline{x}(k-1)\cdots\overline{x}(k-k+1)]^T$, and $\overline{X}(k) \in R^{\theta}$. Based on Eq. (4.62), it is deduced that:

$$\overline{X}(k+1)=\overline{\Omega'}\overline{X}(k)+\overline{\Psi'}[\overline{U}(k)+\overline{F''}(k)]\tag{4.63}$$

where $\overline{\Omega'} = Diag\{\overline{\Omega}, \overline{\Omega}, ..., \overline{\Omega}\}$, $\overline{\Psi'} = Diag\{\overline{\Psi}, \overline{\Psi}, ..., \overline{\Psi}\}$, $\overline{U}(k) = [u(k)\,u(k-1)\cdots u(k-k+1)]^T$, $\overline{F''}(k) = Diag\{\overline{F'}(k), \overline{F'}(k-1), \cdots, \overline{F'}(k-k+1)\}$, and $Diag\{\cdot\}$ denotes the block diagonal matrix.

Let $rank(\overline{\Omega'}) = l$; there exist a nonsingular matrix $\overline{\Xi}$ so that:

$$\overline{\Xi\Psi'}=\begin{bmatrix}0_{(\overline{\theta}-l)\times l}\\\overline{\Psi'}_2\end{bmatrix}\tag{4.64}$$

where $\overline{\Psi'}_2 \in R^{l\times l}$ is nonsingular.

Let $\overline{Z}(k) = \overline{\Xi X}(k)$; thus, the closed-loop NCS can be expressed as:

$$\overline{Z}(k+1)=\widehat{\overline{\Omega}}\overline{Z}(k)+\begin{bmatrix}0_{(\overline{\theta}-l)\times l}\\\overline{\Psi'}_2\end{bmatrix}[\overline{U}(k)+\overline{F''}(k)]\tag{4.65}$$

where $\widehat{\overline{\Omega}} = \overline{\Xi\Omega'\Xi}^{-1}$.

Furthermore, the closed-loop NCS can be written as

$$\overline{Z}_1(k+1)=\widehat{\overline{\Omega}}_{11}\overline{Z}_1(k)+\widehat{\overline{\Omega}}_{12}\overline{Z}_2(k)\tag{4.66a}$$

$$\overline{Z}_2(k+1)=\widehat{\overline{\Omega}}_{21}\overline{Z}_1(k)+\widehat{\overline{\Omega}}_{22}\overline{Z}_2(k)+\overline{\Psi}'_2[\overline{U}(k)+\overline{F''}(k)]\tag{4.66b}$$

It is evident that the equation of the NCS (4.66a) represents the sliding motion dynamics of NCS (4.65) [27]. The following equalities can be obtained by combining Eqs. (4.7), (4.19), (4.20), and (4.21a):

$$\overline{\sigma}(k) = [1\ \omega_{k-1} \cdots \omega_1]\begin{bmatrix} \overline{S}(k) \\ \overline{S}(k-1) \\ \vdots \\ \overline{S}(k-k+1) \end{bmatrix}, \begin{bmatrix} \overline{S}(k) \\ \overline{S}(k-1) \\ \vdots \\ \overline{S}(1) \end{bmatrix} =$$

$$\begin{bmatrix} \overline{s}_{11} & \cdots & \overline{s}_{1k} \\ \vdots & \vdots & \vdots \\ \overline{s}_{k1} & \cdots & \overline{s}_{kk} \end{bmatrix}\begin{bmatrix} \overline{e}(k) \\ \overline{e}(k-1) \\ \vdots \\ \overline{e}(k+k-1) \end{bmatrix}, \begin{bmatrix} \overline{e}(k) \\ \overline{e}(k-1) \\ \vdots \\ \overline{e}(1) \end{bmatrix} = \begin{bmatrix} \overline{e}_{11} & \cdots & \overline{e}_{1k} \\ \vdots & \vdots & \vdots \\ \overline{e}_{k1} & \cdots & \overline{e}_{kk} \end{bmatrix}\begin{bmatrix} \overline{x}(k) \\ \overline{x}(k-1) \\ \vdots \\ \overline{x}(k-k+1) \end{bmatrix} =$$

$\overline{EX}(k) = \overline{E\Xi}^{-1}\overline{Z}(k)$. Hence, the k-order sliding-mode function can be written as follows:

$$\begin{aligned}\overline{\sigma}(k) &= \overline{\widehat{E}}\overline{Z}(k) = [\overline{\widehat{E}}_1\ \overline{\widehat{E}}_2][\overline{Z}_1(k)\ \overline{Z}_2(k)]^T \\ &= \overline{\widehat{E}}_1\overline{Z}_1(k) + \overline{\widehat{E}}_2\overline{Z}_2(k) = 0\end{aligned} \tag{4.67}$$

where $\overline{\widehat{E}}_1 \in R^{l\times(\overline{\theta}-l)}$. By substituting $\overline{Z}_2(k) = -\overline{\widehat{E}}_2^{-1}\overline{\widehat{E}}_1\overline{Z}_1(k)$ into Eq. (4.66a), we obtain:

$$\overline{Z}_1(k+1) = (\overline{\widehat{\Omega}}_{11} - \overline{\widehat{\Omega}}_{12}\overline{\widehat{E}}_2^{-1}\overline{\widehat{E}}_1)\overline{Z}_1(k) \tag{4.68}$$

Theorem 2 *The reduced-order system* (4.68) *is quadratically stable if there exists a symmetric positive-definite matrix* $\overline{P} \in R^{(\overline{\theta}-l)\times(\overline{\theta}-l)}$ *so that*:

$$\begin{bmatrix} -\overline{P} & (\overline{\widehat{\Omega}}_{11} - \overline{\widehat{\Omega}}_{12}\overline{\widehat{E}}_2^{-1}\overline{\widehat{E}}_1)^T \\ (\overline{\widehat{\Omega}}_{11} - \overline{\widehat{\Omega}}_{12}\overline{\widehat{E}}_2^{-1}\overline{\widehat{E}}_1) & -\overline{P}^{-1} \end{bmatrix} < 0 \tag{4.69}$$

Proof Construct the following Lyapunov function:

$$\overline{V}(k) = \overline{Z}_1(k)^T\overline{PZ}_1(k)$$

The Lyapunov theory states that the difference in $\overline{V}(k)$ along the trajectory of the system (4.68) yields:

$$\begin{aligned}\Delta\overline{V}(k) &= \overline{Z}_1^T(k+1)\overline{PZ}_1(k+1) - \overline{Z}_1^T(k)\overline{PZ}_1(k) \\ &= \overline{Z}_1^T(k)(\overline{\widehat{\Omega}}_{11} - \overline{\widehat{\Omega}}_{12}\overline{\widehat{E}}_2^{-1}\overline{\widehat{E}}_1)^T\overline{P}(\overline{\widehat{\Omega}}_{11} - \overline{\widehat{\Omega}}_{12}\overline{\widehat{E}}_2^{-1}\overline{\widehat{E}}_1)\overline{Z}_1(k) - \overline{Z}_1^T(k)\overline{PZ}_1(k) < 0\end{aligned} \tag{4.70}$$

Hence:

$$(\overline{\overset{\frown}{\Omega}}_{11} - \overline{\overset{\frown}{\Omega}}_{12}\overline{\overset{\frown}{E}}_2^{-1}\overline{\overset{\frown}{E}}_1)^T\overline{P}(\overline{\overset{\frown}{\Omega}}_{11} - \overline{\overset{\frown}{\Omega}}_{12}\overline{\overset{\frown}{E}}_2^{-1}\overline{\overset{\frown}{E}}_1) - \overline{P} < 0 \tag{4.71}$$

By using the Schur complement, the matrix inequality (4.69) can be obtained. The proof is completed.

4.5 Simulation Results

The simulation is divided into two parts: the 2nd-order sliding mode controller and the k-order one.

4.5.1 Simulation Results of 2-order Sliding Mode Control Method

The DC motor can be formulated as follows:

$$\begin{cases} \dot{\overline{x}}(t) = \begin{bmatrix} 1 & 0.001 \\ 0 & 0.9753 \end{bmatrix}\overline{x}(t) + \begin{bmatrix} 0.0001 \\ 0.1314 \end{bmatrix}u(t) \\ y(t) = [\,1\ 0\,]\overline{x}(t) \end{cases}$$

In addition, the disturbance is $\overline{F}(k) = \overline{G}_1 \times rand1 \times \sin(8k\pi) + \overline{G}_2 \times rand2 \times \cos(10k\pi)$, where $\overline{G}_1$ and $\overline{G}_2$ are the gain parameters of the disturbance, and $rand1, rand2 \in (0, 1)$ are random numbers. The reference input is $ref(k) = 6\sin(8\pi k) + 4\cos(10\pi k)$.

The simulation results consider the following two cases.

Case 1: Constant Delay

The constant delay is considered in the communication network. The initial values of the system state and output are $\overline{x}_0 = [00]^{\mathrm{T}}$ and $y_0 = 0$. The controller parameters are $\overline{\beta} = 1.1, \overline{\gamma} = 0.45, \overline{h} = 0.001$, and $\overline{\alpha} = 0.6$. The disturbance gains are $\overline{G}_1 = 1.5$ and $\overline{G}_2 = 2$. Let the desired poles of the observer be [0.10.3]. By using the pole assignment method, the following observer gain is obtained:

$$\overline{L} = [\,1.5753,\ 591.0901]^{\mathrm{T}}$$

The simulation results are shown in Figs. 4.4 and 4.5. Figure 4.4 depicts the input perturbations of the system. When the network is in an ideal state (i.e. $RTT = 0$), as shown in Fig. 4.5, the system is only affected by perturbations. It can be seen that

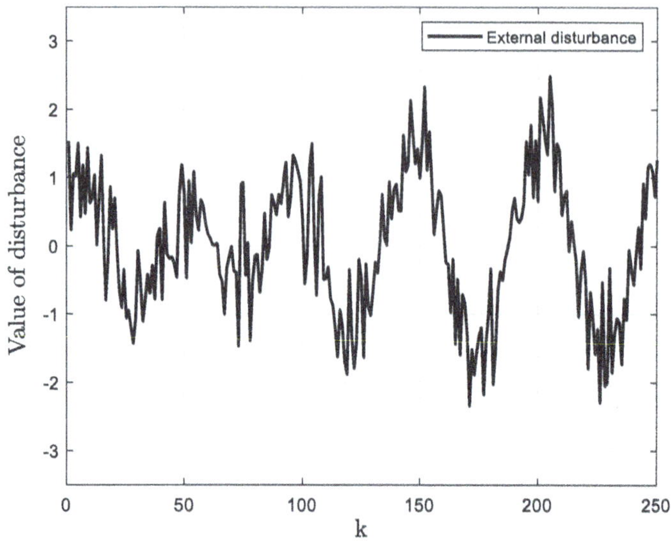

Fig. 4.4. Input disturbance

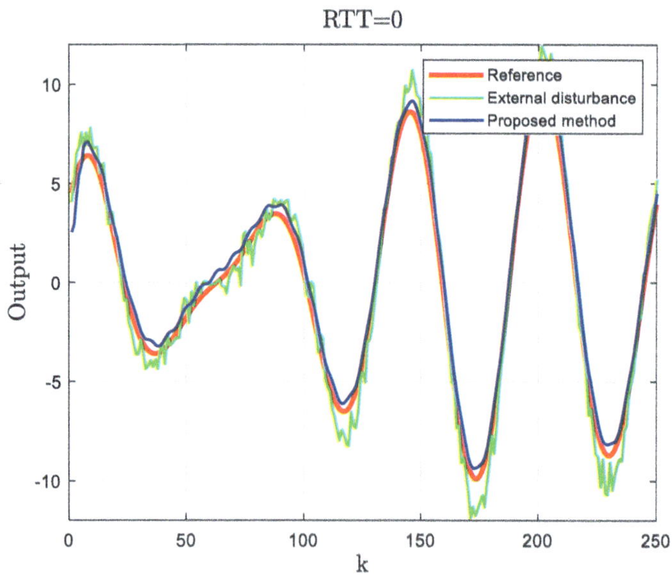

Fig. 4.5 Tracking output of the system under $RTT = 0$

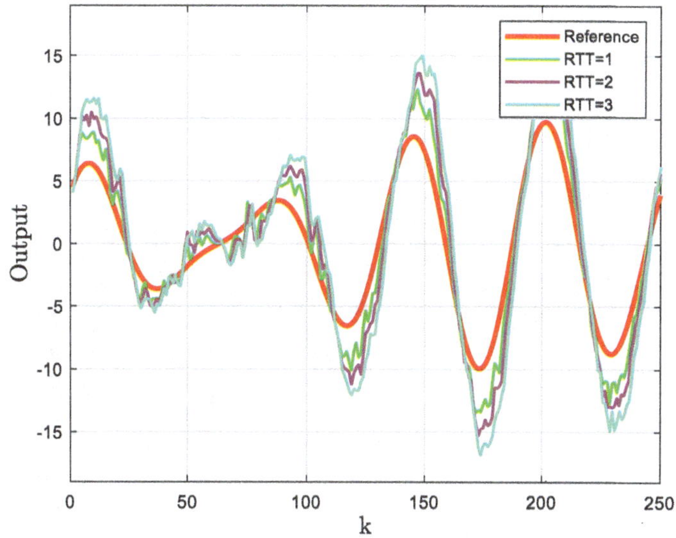

Fig. 4.6 Tracking output of the system under different delay

the proposed method effectively suppresses the input disturbance. Figure 4.6 shows the tracing output of the system with different delays. The output trace performance degrades as the latency increases.

Figure 4.7 depicts the tracing output of the system using the proposed method for $RTT = 2$. It can be seen that the proposed method has better control performance than the method with no compensation. The simulation results show that the proposed method successfully suppresses the disturbance and mitigates the delay.

Case 2: Random Delay

In contrast to case 1, in case 2, the delay RTT is random, i.e.,$RTT \in \{0, 1, 2, 3\}$ The gains of the disturbance and sampling period are the same as in case 1. The parameters of the controller are $\overline{\beta} = 0.8, \overline{\gamma} = 0.1$, $\overline{\alpha} = 0.55$, and $\overline{\kappa} = 0.06$. The desired poles of the observer are [0.15 0.4]. By using the pole assignment method, the following observer gain is obtained:

$$\overline{L} = [1.4253,\ 474.7951]^{\mathrm{T}}$$

Figure 4.8 shows the input perturbations in the system. The random delay in the network is shown in Fig. 4.9. It can be seen that the random delay RTT has the values $\{0, 1, 2, 3\}$. Figure 4.10 depicts the tracking output of the system with random delay. In the absence of compensation, the tracking output curve exhibits significant jitter. In addition, the control performance is improved substantially after using the proposed method. The simulation results show that the proposed method effectively suppresses the input disturbance and compensates for random delay.

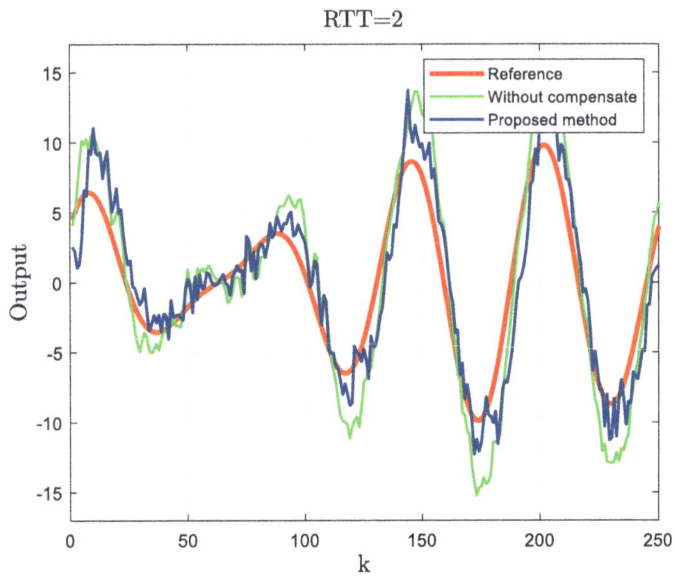

Fig. 4.7 Tracking output of system using proposed method under $RTT = 2$

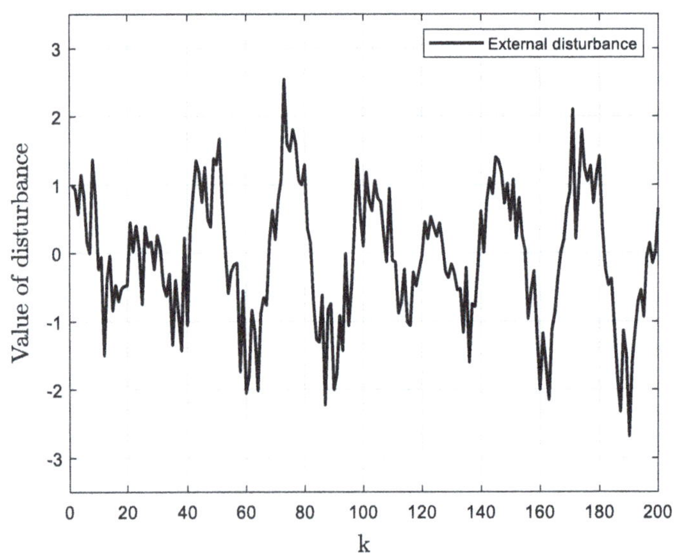

Fig. 4.8 Input disturbance

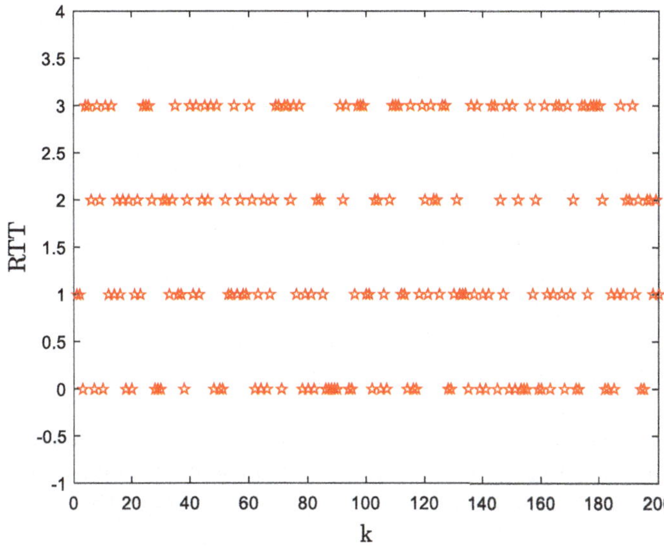

Fig. 4.9 Random delay in network (0–3 RTT)

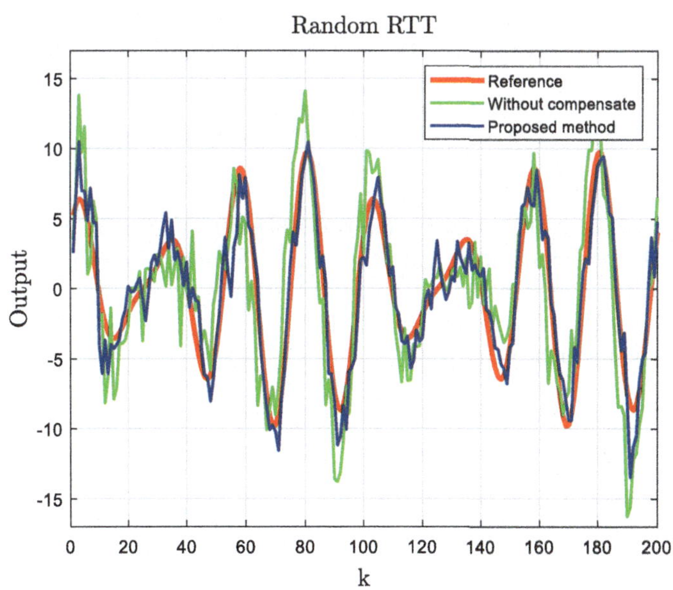

Fig. 4.10. Tracking output of system using proposed method under random delay

4.5.2 *Simulation Results of k-order Sliding Mode Control Method*

In this subsection, a servo DC motor system is considered in the simulations to demonstrate the effectiveness of the proposed method. The differential equation is designed as in Ref. [26]:

$$\frac{d\bar{i}_a(t)}{dt} = -\frac{\overline{R}_a}{\overline{L}_a}\bar{i}_a(t) - \frac{\overline{K}_b}{\overline{L}_a}\omega(t) + \frac{1}{\overline{L}_a}\overline{V}_a(t)$$
$$\frac{d\overline{\omega}(t)}{dt} = -\frac{\overline{B'}}{\overline{J}}\omega(t) + \frac{\overline{K}_T}{\overline{J}}\bar{i}_a(t)$$

where $\overline{\mathrm{V}}_a(t)$ is the armature voltage, $\bar{i}_a(t)$ represents the armature current, $\overline{V}_a(t)$ is the resistance, and $\overline{L}_a$ is the inductance of the armature winding in the motor. $\overline{B'}$ is the damping in the motor and the load of the fixed chassis, and $\overline{\omega}(t)$ is the angular velocity.

Let $\overline{x} = \begin{pmatrix} \overline{x}_1 \\ \overline{x}_2 \end{pmatrix} = \begin{pmatrix} \bar{i}_a \\ \overline{\omega} \end{pmatrix}$, $u = \overline{V}_a$, and $y = \overline{\upsilon} = \frac{30}{\pi}\overline{x}_2$, where $\overline{\upsilon}$ represents the speed of the motor. $\overline{x}$, u, and y are the state, control input, and output of the system.

According to Eqs. (4.1) and (4.2), the following formulations can be obtained:

$$\overline{A} = \begin{bmatrix} -\frac{\overline{R}_a}{\overline{L}_a} & -\frac{\overline{K}_b}{\overline{L}_a} \\ \frac{\overline{K}_T}{\overline{J}} & -\frac{\overline{B'}}{\overline{J}} \end{bmatrix}, \overline{B} = \begin{bmatrix} 1/\overline{L}_a \\ 0 \end{bmatrix} \text{ and } \overline{\mathrm{C}} = \begin{bmatrix} 0 \ 30/\pi \end{bmatrix}.$$

The type of servo DC motor system is GNM5440, which is selected following Ref. [26]. The nominal voltage and power are 24 V and 95 W, respectively. $\overline{R}_a = 0.965(Ohms)$, $\overline{L}_a = 0.0022\,(H)$, $\overline{K}_b = 0.1201(V \cdot s/rad)$, $\overline{K}_T = 0.1201(N \cdot m/A)$, $\overline{B'} = 0.1296(N/ms)$, and $\overline{J} = 0.1182(kgm^2)$.

The following can be derived:

$$\overline{A} = \begin{bmatrix} -438.6364 & -54.5909 \\ 1.0161 & -1.0964 \end{bmatrix}, \overline{B} = \begin{bmatrix} 454.5455 \\ 0 \end{bmatrix} \text{ and } \overline{\mathrm{C}} = \begin{bmatrix} 0 \ 9.5493 \end{bmatrix}.$$

We use the discretization method to obtain the discrete-time model of the servo system:

$$\overline{\Omega} = \begin{bmatrix} -0.0003 & -0.1189 \\ 0.0022 & 0.9525 \end{bmatrix}, \overline{\Psi} = \begin{bmatrix} 1.0317 \ 0.0388 \end{bmatrix}^{\mathrm{T}} \text{ and } \overline{\Gamma} = \begin{bmatrix} 0 \ 9.5493 \end{bmatrix}$$

The external disturbance is defined as follows:

$$\overline{F}(k) = \overline{G}_1 \times rand1 \times \sin(7k\pi) + \overline{G}_2 \times rand2 \times \cos(2k\pi)$$

where $\overline{G}_1$ and $\overline{G}_2$ are the gains of the disturbance, and $rand1, rand2 \in (0, 1)$ describe the random values.

The reference signal is a square wave between −900 and 1000. The sampling period is 0.04 s, and the simulation time is 10 s. For comparison, the 2-DTSMC strategy used in [23] is evaluated. The following two cases are considered.

Case 1: constant RTT delay

Constant RTT delay is introduced into the communication network. The initial values of the system are as follows: $\overline{x}_0 = [00]^{\mathrm{T}}$, $\overline{y}_0 = 0$. The controller parameters are $\overline{\beta} = 1.5$, $\overline{\gamma} = 0.15$, $\overline{\alpha} = 0.8$, and $\overline{\kappa} = 0.1$. The disturbance gains are $\overline{G}_1 = 3$ and $\overline{G}_2 = 5$. The observer gain is $\overline{L} = [19841, \ -46]^{\mathrm{T}}$.

Figure 4.11 represents the external disturbances introduced into the system. It is a sine wave signal with a random superposition of two amplitudes.

In Fig. 4.12, the tracking output of the system for different RTT is shown. The RTT ranges from 1 to 3. The simulation results indicate that the tracking error increases as the RTT delay increases. In Fig. 4.13, the RTT delay is 3. The delay and packet loss are compensated by the k-ADSMC and 2-DTSMC described in [23]. The response and tracking performance of the system are faster and better for the proposed k-ADSMC method than the 2-DTSMC method proposed in [23].

Case 2: random RTT delay

In this case, the RTT delay is random in the communication network. The disturbances gain parameters are $\overline{G}_1 = 3$ and $\overline{G}_2 = 5$. The controller parameters are $\overline{\beta} = 1.2$,$\overline{\gamma} = 0.5$,$\overline{\alpha} = 0.9$, and $\overline{\kappa} = 0.01$, and the observer gain is $\overline{L} = [19832, \ -46]^{\mathrm{T}}$.

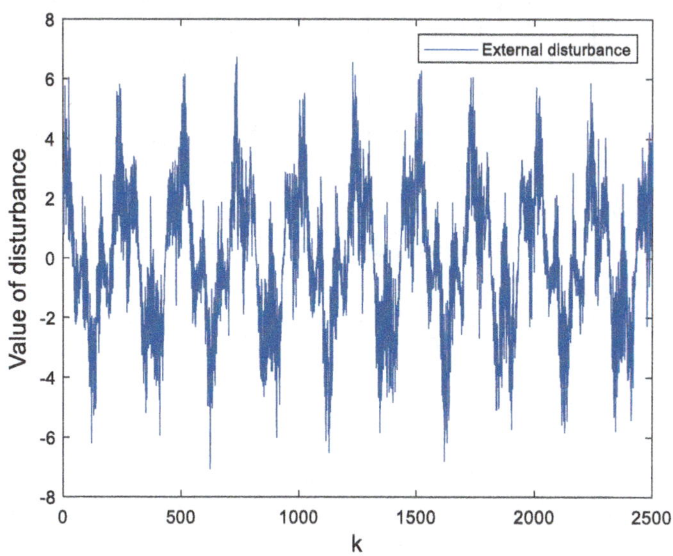

Fig. 4.11 External disturbance

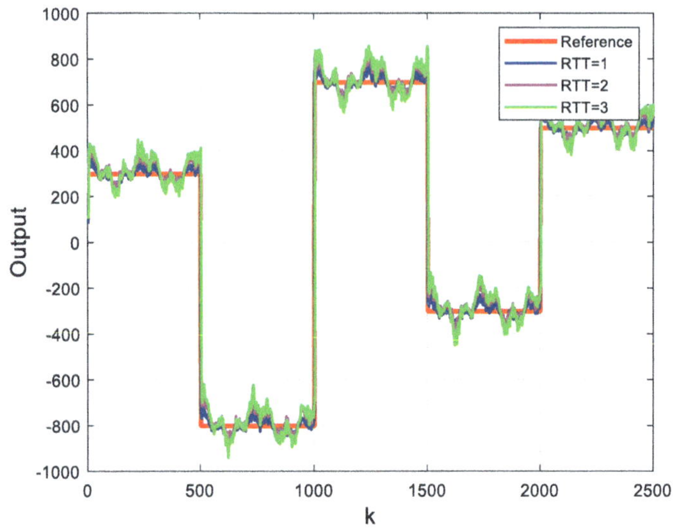

Fig. 4.12 Tracking output of the system under different RTT

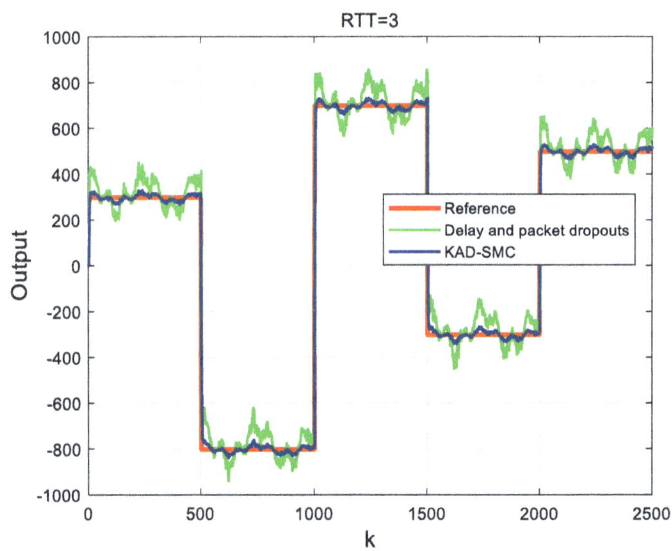

Fig. 4.13 Tracking output of system when RTT set as 3

Figure 4.14 shows the external disturbances introduced into the system using random RTT delays. It is a sine wave signal with a random superposition of two amplitudes.

The random RTT delay of steps 1–5 is shown in Fig. 4.15. The RTT delay is randomly distributed between 1 and 5. As shown in Fig. 4.16, the tracking output of

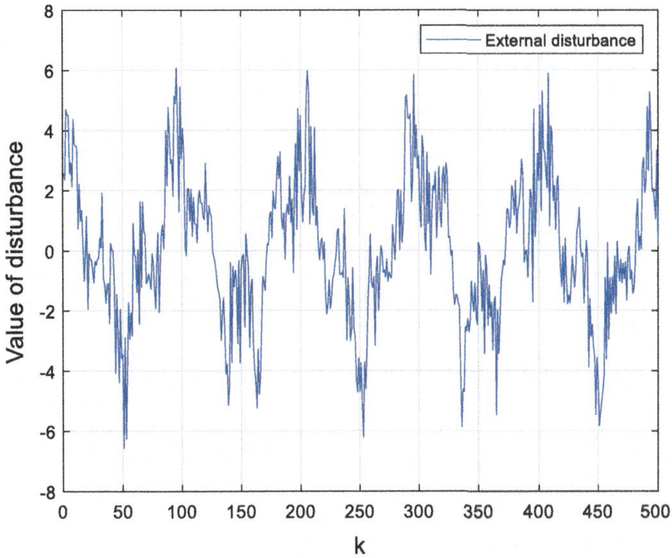

Fig. 4.14 External disturbances of the system

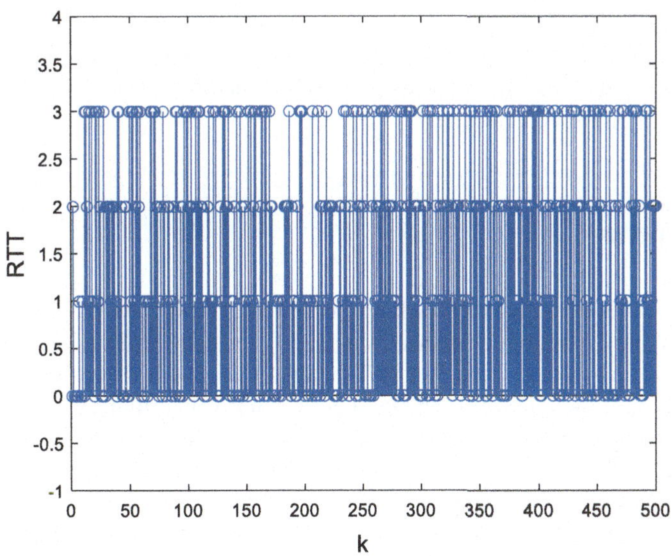

Fig. 4.15 Random RTT delays

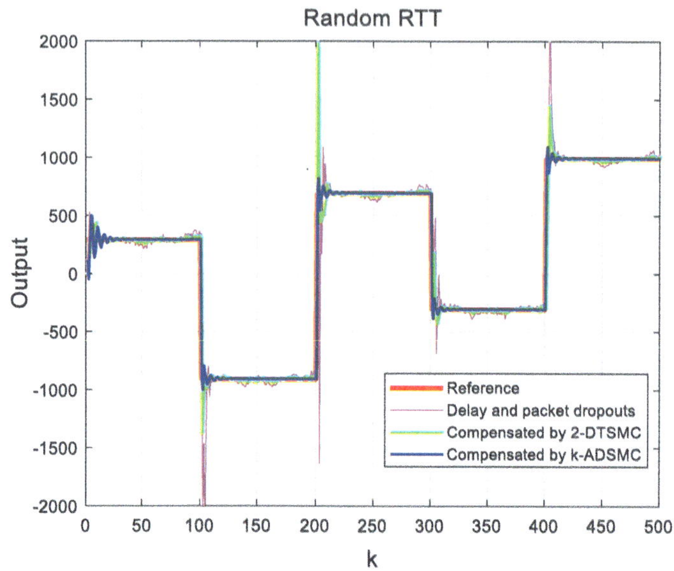

Fig. 4.16 Tracking performance under different compensation method

k-ADSMC remains stable, providing good tracking performance. In addition, unlike the 2-DTSMC method, the proposed k-ADSMC method substantially reduces the jitter as the motor speed changes.

4.6 Conclusions

This chapter investigated the tracking control of an NCS with communication constraints and external interference, considering the delay, packet loss, and packet disorder caused by the network. The RTT delay was used to describe the communication constraints. The 2-DSMTC and k-ADSMC control algorithms were proposed to overcome communication constraints and external interference. A new DSM function based on the tracking error and error difference was used in the 2-DSMTC. In addition, the 2-DSMTC method included a disturbance estimate as the weight factor. Finally, an NPC algorithm based on 2-DSMTC was proposed. The simulation results demonstrated the effectiveness of the proposed method. In the k-ADSMC, an adaptive DSM function was established to adjust the sliding mode parameters dynamically according to the current state of the system. In addition, a k-th order sliding mode function was designed to consider external interference. This function used all k the second estimate of the disturbance and the sliding mode value to estimate the influence of external disturbance and improve the tracking performance. Then, an NPC method based on k-ADSMC was proposed to compensate for the communication constraints. Finally, a DC motor system was used as an example to verify the effectiveness of the method. The experimental results showed that the proposed k-ADSMC method remained stable and showed good tracking performance regardless of whether constant or random RTT delays occurred.

References

1. M. Li and Y. Chen, "Robust Tracking Control of Networked Control Systems with Communication Constraints and External Disturbance," IEEE Transactions on Industrial Electronics, vol. 64, no. 5, pp. 4037-4047, May 2017.
2. S. Wang, X. Meng, and T. Chen, "Wide-area control of power systems through delayed network communication," IEEE Trans. Control Syst. Technol., vol.20, no.2, pp. 495-503, Mar. 2012.
3. J. Emil, B. Raquel, P. Munoz, I. D. L. Bandera, and I. Serrano, "Self-healing in mobile networks with big data," IEEE Commu. Mag., vol. 54, no. 1, pp. 114-120, Jan. 2016.
4. Y. Xia, "Cloud control systems," IEEE/CAA J. Autom. Sinica, vol. 2, no. 2, pp. 134-142, Apr. 2015
5. M. Cho and C. Lee, "A low-power real-time operating system for ARC (actual remote control) wearable device," IEEE Trans. Consum. Electron., vol. 56, no.3, pp. 1602-1609, Aug. 2010.
6. W. Heemels, A. Teel, N. N. D. Wouw and D. Nešíc, "Networked control systems with communication constraints: Tradeoffs between transmission intervals, delays and performance," IEEE Trans. Autom. Control, vol. 55, no. 8, pp. 1781–1796, Aug. 2010.
7. G. Franzè, F. Tedesco, and D. Famularo, "Model predictive control for constrained networked systems subject to data losses," Automatica, vol. 54, pp. 272–278, Apr. 2015.
8. R. Wang, B. Wang, G. Liu, W. Wang, and D. Rees, "H∞ Controller design for networked predictive control systems based on the average dwell-time approach," IEEE Trans. Circuits Syst. II, Exp. Briefs, vol. 57, no. 4, pp. 310- 314, Apr. 2010.
9. Y. Liu, and I. Fong, "Robust predictive tracking control of networked control systems with time-varying delays and data dropouts," IET Control Theory Appl., vol. 7, no.5, pp. 738–748, May. 2013.
10. R. Yang, G. Liu, P. Shi, C. Thomas, and M. V. Basin, "Predictive output feedback control for networked control systems," IEEE Trans. Ind. Electron., vol. 61, no. 1, pp. 512- 520, Jan. 2014.
11. J. Zhang, Y. Xia, and P. Shi, "Design and stability analysis of networked predictive control systems," IEEE Trans. Control Syst. Technol., vol. 21, no. 4, pp. 1495-1501, Jul. 2013.
12. Z. Pang, and G. Liu, "Design and implementation of secure networked predictive control systems under deception attacks," IEEE Trans. Control Syst. Technol., vol. 20, no. 5, pp. 1334-1342, Sep. 2012.
13. Y. Shi, J. Huang, and B. Yu, "Robust tracking control of networked control systems: application to a networked dc motor," IEEE Trans. Ind. Electron., vol. 60, no. 12, pp. 5864- 5874, Dec. 2013
14. H. Zhang, Y. Shi, and M. Liu, "H∞ step tracking control for networked discrete-time nonlinear systems with integral and predictive actions," IEEE Trans. Ind. Informat., vol. 9, no. 1, pp. 337-345, Feb. 2013
15. H. Gao, and T. Chen, "Networked-based H∞ output tacking control," IEEE Trans. Autom. Control, vol. 53, no. 9, pp. 2142–2148, Oct. 2008
16. Y. Wang, and G. Yang, "Output tracking control for networked control systems with time delay and packet dropout," Int. J. Control, vol. 81, no. 11, pp. 1709–1719, Nov. 2008
17. Y. Wang, and G. Yang, "Robust H∞ model reference tracking control for networked control systems with communication constraints," Int. J. Control Autom. Syst., vol. 7, no. 6, pp. 992-1000, Dec. 2009.
18. Z. Pang, G. Liu, D. Zhou, and M. Chen, "Output tracking control for networked systems: a model-based prediction approach," IEEE Trans. Ind. Electron., vol. 61, no. 9, pp. 4867- 4877, Sep. 2014.

19. C. Zhou, Y. Shi, S. Yang, Q. Yin, and Y. Qin, "Characteristic model-based adaptive discrete-time sliding mode control for the swing arm in a fourier transform spectrometer," IEEE Trans. Sys., Man, Cybern., vol. 42, no. 6, pp. 1633- 1643, Nov. 2012.
20. X. Zhang, L. Sun, K. Zhao, and L. Sun, "Nonlinear speed control for PMSM system using sliding-mode control and disturbance compensation techniques," IEEE Trans. Ind. Electron., vol. 28, no. 3, pp. 1358- 1365, Mar. 2013
21. H. Du, X. Yu, M. Chen, and S. Li, "Chattering-free discrete-time sliding mode control," Automatica, vol. 68, pp. 87–91, Jun. 2016
22. B. Veselí, B. P. Draženović, and C. Milosavljevic, "Improved discrete-time sliding-mode position control using euler velocity estimation," IEEE Trans. Ind. Electron., vol. 57, no. 11, pp. 3840- 3847, Nov. 2010
23. Q. Xu, "Piezoelectric nanopositioning control using second-order discrete-time terminal sliding-mode strategy," IEEE Trans. Ind. Electron., vol. 62, no. 12, pp. 7738- 7748, Dec. 2015
24. W. Su, S. V. Drakunov, and U. Ozguner, "An O(T 2) boundary layer in sliding mode for sampled-data systems," IEEE Trans. Autom. Control, vol. 45, no. 3, pp. 482–485, May. 2000.
25. Y. A. Alma, A. D. Nancy, and L. F. Carlos, "Bacterial foraging optimization algorithm to improve a discrete-time neural second order sliding mode controller," Appl. Math. Comput., vol. 271, no. 15, pp. 43–51, Nov. 2015.
26. S. O. Ramon, M. H. G. Victor, A. C. Mayra, and M. C. Daniel, "DC/DC buck power converter as a smooth starter for a dc motor based on a hierarchical control," IEEE Trans. Power Electron., vol. 30, no. 2, pp. 1076-1084, Feb. 2015.
27. H. Li, H. Yang, F. Sun, and Y. Xia, "Sliding-mode predictive control of networked control systems under a multiple-packet transmission policy," IEEE Trans. Ind. Electron., vol. 61, no. 11, pp. 6234- 6243, Nov. 2014

Chapter 5
Fast Sliding-Mode Tracking Control for Networked Systems with Actuator Faults

In this chapter, faults and disturbances affecting actuators and sensors in networked control systems (NCSs) are considered. A fast high-order sliding mode (FHOSM) controller is proposed to compensate for actuator failure and sensor interference. The proposed controller is constructed based on the estimated information of the NCS. Therefore, adaptive observers with multiple levels are designed to provide state and sensor interference estimates and protect the NCS from actuator failures. In addition, a new sliding function is proposed to achieve finite-time convergence. The stability of the system with the applied control is analyzed. A simulation of an NCS with faults and interference is carried out. The results show that the proposed method is effective for preventing the two problems of the system and suppressing failure.

5.1 Introduction

Due to developments in communication and information technology, NCSs are used extensively. An NCS is used to distribute and share data between different applications. However, NCSs face significant challenges due to system failures, external interference, sensor noise, network-induced latency, packet loss, and hacking, which are common in actuator components. Faults and interference may affect the reliability of the entire NCS or cause instability of the system. These problems may reduce the operational ability of the NCS tracking system [1, 2]. In addition, due to rotating parts and operating conditions in many applications, such as unmanned aerial vehicles (UAVs), autonomous vehicle systems, electric vehicles, and aerospace applications, vulnerability to failures and interference are crucial issues [3, 4].

Figure 5.1 shows the configuration of a control system with actuator failure and sensor interference; the sensors and actuators are mounted on the plant side. Sensors are installed to monitor the system's variable parameters, such as the current, voltage, and angular velocity. The data are transmitted to the remote control center through a wired/wireless network feedback channel. The remote control center computes the

© The Author(s), under exclusive license to Springer Nature Singapore Pte Ltd. 2022
M. Li et al., *Tracking Control of Networked Systems via Sliding-Mode*,
https://doi.org/10.1007/978-981-16-6514-1_5

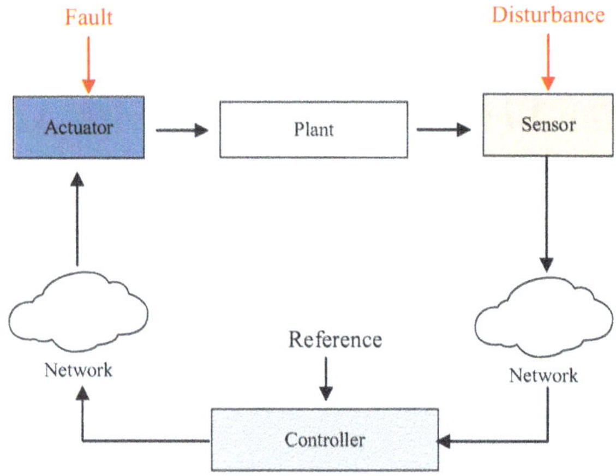

Fig. 5.1 Block diagram of wired/wireless control system under disturbances and faults

input information and sends it through the feedforward channel to the actuator, which uses the input data to drive the system. Failures adversely affect the actuators and create interference in the actuators and sensors. As a result, the NCS will lose the ability to link the system components, compromising the system.

Fault-tolerant control (FTC) has been widely used to reduce risks due to faults and disturbances in NCSs. Many FTC techniques have been proposed and provided acceptable system performance and improved system reliability [5–15]. The authors in [5, 6] investigated the FTC of automated systems against unknown disturbances and system failures. In [7], data-driven output-feedback FTC in unknown systems with random measurements and process noise was analyzed. In [8], an integrated sliding mode fault-tolerant controller was designed to ensure good control performance in uncertain network systems. In [9], an integrated FTC technique based on linear matrix inequality (LMI) was designed to mitigate the influence of faults, ensure the stability of the input, and ensure the robustness of linear and nonlinear systems. Enhanced-state unknown input observer and high-order sliding mode observation techniques were proposed in [10, 11] to obtain fault estimation (FE) information and compensate for faults, respectively. However, the above method is also effective for compensating for faults under very strict matching conditions, which limits the controller attributes, such as the design degrees of freedom and the time convergence of the errors. Recently, actuator failures and sensor failures in NCSs have attracted researchers' attention [12, 13]. Nonetheless, actuator and sensor failures remain challenging issues to date, and research is ongoing. References [14, 15] described small studies concern the the evolution of faults over time. However, the authors focused on the worst-case design conditions and placed greater emphasis on handling failures to keep the system running, regardless of whether there are different failure conditions. They did not consider general failures and their economic impacts.

Due to its nonlinear behavior, sliding mode control (SMC) ensures convenient control performance despite external disturbances, sensor/actuator failures, and disturbances to the system [16–18]. Reference [19] designed a K-order adaptive sliding mode controller with good interference suppression. However, the controller caused a large overshoot, resulting in an overflow of the system. Moreover, the SMC had a vibration problem, which significantly affected its application. In recent years, many successful methods have been proposed to improve jitter. For example, the high-order sliding mode [20], super-twisting algorithm (STW) [21–24], and adaptive techniques [25] were proposed. The authors proposed the reaching law, which proved effective for alleviating jitter [26]. In reference [27], a novel nonsingular fast terminal SMC with finite time tolerance for passive and active systems was introduced to compensate for problems in uncertain systems. Reference [28] proposed an improved discrete arrival law in the quasi-sliding mode domain for discrete-time systems with uncertainties.

The contributions of this paper include the following to overcome the limitations of existing FTC methods [14, 15, 27]. First, a variable structure multi-level observer is designed to estimate and protect the NCS against different actuator fault types. Second, an FHOSM controller based on the hypertwist algorithm is constructed to compensate for disturbances of the actuator and sensor and achieve finite-time convergence without jitter. In addition, the system dynamics of the proposed control law are analyzed, and the stability of the proposed algorithm is demonstrated. Finally, the control algorithm is verified using a networked DC motor control system, and the results are compared with the pushback control method described in [15]. The proposed method has advantages over the FE, FTC, and integrated FE/FT methods. The simulation results show that this method has advantages over other methods for compensating faults and disturbances of the actuator and sensor channels. The rest of this article is organized as follows. The second section describes the fault and interference problems in the actuator and sensor channels of the NCS.

The proposed control system design is presented in the third section. A multi-stage observer is designed, and the FHOSM controller is constructed based on the estimation information of the NCS, demonstrating the convergence of the proposed algorithm. In the fourth section, a numerical simulation is carried out to prove the effectiveness of the proposed algorithm. Finally, Section V provides the conclusion.

5.2 Problem Formulation

Figure 5.2. shows the NCS structure affected by faults and disturbances, where $\overline{\Phi}$ and $\overline{\Psi}$ are the actuator fault and sensor disturbances, respectively. The observer includes a virtual switch to obtain estimation information based on different fault conditions. The proposed new functionality works as a virtual switch.

The characteristics of a single-input single-output (SISO) system under actuator failure and sensor interference are (5.1):

$$\begin{cases} \dot{\overline{\chi}}(t) = \overline{A}\overline{\chi}(t) + \overline{B}u(t) + \overline{\alpha}\overline{\Phi} + \overline{\beta}_a\overline{\Psi}(t) \\ \overline{\Upsilon}(t) = \overline{C}\overline{\chi}(t) + \overline{\beta}_s\overline{\Psi}(t) \end{cases} \tag{5.1}$$

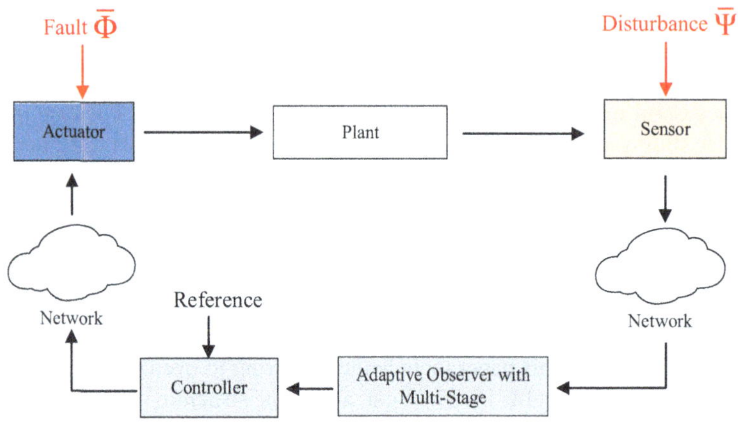

Fig. 5.2 Proposed block diagram of networked system under disturbances and faults

where $\overline{\chi}(t) \in R^n$ is the state vector, $u(t) \in R^m$ is the control input, $\overline{y}(t) \in R^p$ is the control output, $m = p = 1$, $\overline{\Phi} \in R^l$ is the fault signal in the actuator, and $\overline{\Psi} \in R^p$ represents the disturbance affecting the sensor channels. Before designing the defense strategy, the following variables are defined in references to [15]:

Let $\tilde{\overline{\chi}} = [\overline{\chi}^T\ \overline{\Psi}]^T$, $\overline{\Xi} = [I_n\ O_{nxq}]$, $\tilde{\overline{A}} = [\overline{A}\ \overline{\beta}_a]$ and $\tilde{\overline{C}} = [\overline{C}\ \overline{\beta}_s]$. Accordingly, the new extended system can be formulated as follows:

$$\overline{\Xi}\dot{\tilde{\overline{\chi}}} = \tilde{\overline{A}}\tilde{\overline{\chi}} + \overline{B}u + \overline{\alpha}\overline{\Phi}$$
$$\overline{\Upsilon} = \tilde{\overline{C}}\tilde{\overline{\chi}}. \tag{5.2}$$

In the following sections, the NCS system defined in Eq. (5.2) is considered a case study.

5.3 Design of Control Methodology

In this section, an adaptive sliding mode multistage observer (ASMMSO) is designed to provide the information estimation required by the controller to compensate for the effects of faults and disturbances in the actuators and sensors.

5.3.1 Design of Observer

The observer of the system is designed as follows:

$$\begin{cases} \dot{\overline{\Omega}} = \overline{M}\overline{\Omega} + \overline{\varphi}u + \overline{H}\Upsilon + \overline{\Gamma}\overline{\varpi}\upsilon \\ \hat{\tilde{\chi}} = \overline{\Omega} + \overline{\delta}\Upsilon \\ \hat{\Upsilon} = \overline{C}\hat{\tilde{\chi}}. \end{cases} \tag{5.3}$$

where $\overline{\Omega} \in R^{n+q}$ denotes the virtual observer state, $\hat{\tilde{\chi}} \in R^{n+q}$ is the system state estimation, υ is an auxiliary switching variable designed to compensate for the effects of the fault $\overline{\Phi}$ in the NCS (5.2). $\overline{M} \in R^{(n+q)\times(n+q)}$, $\overline{\varphi} \in R^{(n+q)\times m}$, $\overline{H} \in R^{(n+q)\times p}$, $\overline{\varpi} \in R^{(n+q)\times p}$, and $\overline{\delta} \in R^{(n+q)\times p}$ are observer parameters to be determined so that $\hat{\tilde{\chi}}$ can converge to $\tilde{\chi}$. $\overline{\Gamma}_i(.)$ is a non-linear adaptive function to consider the fault evolution over time so that:

$$\overline{\Gamma}_i(.) = \begin{cases} \approx 0 & t \le T_c \\ 1 - e^{-k}(t - T_c) & t \ge T_c. \end{cases} \tag{5.4}$$

where T_c is the time when the fault was unmatched, $k > 0$ represents the evolution rate of the unknown fault. When the value of k is small, the fault evolves slowly. It can be called an initial fault, and it ranks within the space of scope that the control effort u spans. If the value of k is large, the function Γ_i approximates a step function so that the discontinuous control variable υ is fully operational. This occurs when a fault is outside the scope covered by the control variable u.

Moreover, let $\overline{\Im} = [I_n \;\; -\overline{C}]^T$ and $\overline{L} = [0_{n\times p}\; I_p]^T$; the following assumption is satisfied:

Assumption 1 The pair $(\overline{\Im}\Xi, \tilde{\overline{C}})$ is observable, that is:

$$\overline{\Im} = \begin{bmatrix} \tilde{\overline{\Xi}} \\ \tilde{\overline{C}} \end{bmatrix}^{-1} \begin{bmatrix} I_n \\ 0_{p\times n} \end{bmatrix} = \begin{bmatrix} I_n \\ -\overline{C} \end{bmatrix}, \overline{\mathrm{L}} = \begin{bmatrix} \tilde{\overline{\Xi}} \\ \tilde{\overline{C}} \end{bmatrix}^{-1} \begin{bmatrix} 0_{n\times p} \\ I_n \end{bmatrix} = \begin{bmatrix} 0_{n\times p} \\ I_p \end{bmatrix}$$

The following example shows when this assumption is satisfied.

$$rank \begin{bmatrix} \tilde{\overline{\Xi}} \\ \tilde{\overline{C}} \end{bmatrix}^{-1} = rank \begin{bmatrix} I_n \; 0_{n\times p} \\ \overline{C}_i \; I_p \end{bmatrix} = n + p \Rightarrow \begin{bmatrix} \tilde{\overline{\Xi}} \\ \tilde{\overline{C}} \end{bmatrix}^{-1} \text{ is invertible.}$$

Definition 1 The faults $\overline{\Phi}$ and perturbations $\overline{\Psi}$ referring to the controller effort u can be categorized as follows:

(1) They are matched if they are inside the range of u, i.e., $rank(\overline{B}, \overline{\alpha}) = rank(\overline{B})$.
(2) They are unmatched if they are outside the range of u, i.e., $rank(\overline{B}, \overline{\alpha}) \neq rank(\overline{B})$.

Theorem 1 is provided to solve the observer parameters:

Theorem 1 *For an NCS that includes the factory defined by Eq. (5.2) and the observer defined by Eq. (5.3), the estimation error* $\overline{e} = \tilde{\chi} - \hat{\tilde{\chi}}$ *converges asymptotically to zero according to the implementation of the adaptive function* (5.4) *and*

Assumption 1. *In addition, the observer parameter matrix* $\overline{\Omega}, \overline{\varphi}, \overline{H}, \overline{\delta}$, *and* $\overline{\varpi}$ *can be obtained for all different fault cases.*

Proof The proof can be divided into the following cases:

Case i. $\overline{\Gamma}_i(.) \approx 0$

In this case, when the fault matches the control input u and $\overline{\alpha} = \overline{a}_1\overline{B}$, the augmented system defined by Eq. (5.2) can be expressed by Eq. (5.5):

$$\begin{aligned}\overline{\Xi}\dot{\tilde{\overline{\chi}}} &= \tilde{\overline{A}}\tilde{\overline{\chi}} + \overline{B}(u + \overline{a}_1\overline{\Phi}) \\ \overline{\Upsilon} &= \tilde{\overline{C}}\tilde{\overline{\chi}}.\end{aligned} \tag{5.5}$$

where $\overline{a}_1$ is a constant. Let $u'' = (u + \overline{a}_1\overline{\Phi})$; thus, the augmented system (5.5) can be written as follows:

$$\begin{aligned}\overline{\Xi}\dot{\tilde{\overline{\chi}}} &= \tilde{\overline{A}}\tilde{\overline{\chi}} + \overline{B}u'' \\ \overline{\Upsilon} &= \tilde{\overline{C}}\tilde{\overline{\chi}}.\end{aligned} \tag{5.6}$$

A right multiplication of both sides of the first equation of Eq. (5.6) with a matrix $\overline{\Im} \in \Re^{(n+q)\times n}$ yields:

$$\overline{\Im\Xi}\dot{\tilde{\overline{\chi}}} = \overline{\Im\tilde{A}}\tilde{\overline{\chi}} + \overline{\Im B}u''. \tag{5.7}$$

By adding $\overline{L\Upsilon}$ to both sides of Eq. (5.7), the following equation can be obtained:

$$\dot{\tilde{\overline{\chi}}} = \overline{\Im\tilde{A}}\tilde{\overline{\chi}} + \overline{\Im B}u'' + \overline{L\Upsilon} \tag{5.8}$$

We denote $u''' = \overline{\Im B}u'' + \overline{L\Upsilon}$ to express the input signal of system (5.8); thus, Eq. (5.9) can be derived:

$$\begin{aligned}\dot{\tilde{\overline{\chi}}} &= \overline{\Im\tilde{A}}\tilde{\overline{\chi}} + u''' \\ \overline{\Upsilon} &= \tilde{\overline{C}}\tilde{\overline{\chi}}\end{aligned} \tag{5.9}$$

According to **Assumption** 1, a Luenberger observer expressed by Eq. (5.10) is designed for the system (5.9):

$$\begin{aligned}\dot{\hat{\tilde{\overline{\chi}}}} &= \overline{\Im\tilde{A}}\hat{\tilde{\overline{\chi}}} + u''' + \overline{S}(\overline{\Upsilon} - \tilde{\overline{C}}\hat{\tilde{\overline{\chi}}}) \\ &= \overline{\Im\tilde{A}}\hat{\tilde{\overline{\chi}}} + \overline{\Im B}u'' + \overline{L\Upsilon} + \overline{S}(\overline{\Upsilon} - \tilde{\overline{C}}\hat{\tilde{\overline{\chi}}})\end{aligned} \tag{5.10}$$

in which $\overline{S}$ is the Luenberger gain matrix. By substituting $\hat{\tilde{\chi}} = \overline{\Omega} + \overline{L\Upsilon}$ into Eq. (5.10), the following is derived:

$$\dot{\overline{\Omega}} = (\overline{\Im\tilde{A}} - \overline{S\tilde{C}})\overline{\Omega} + \overline{\Im B}u'' + (\overline{S} + (\overline{\Im\tilde{A}} - \overline{S\tilde{C}})\overline{L})\overline{\Upsilon} \tag{5.11}$$

By comparing Eq. (5.11) with the observer form in Eq. (5.3), the observer parameters can be solved as:

$$\begin{aligned}
\overline{M} &= \overline{\Im\tilde{A}} - \overline{S\tilde{C}} \\
\overline{\varphi} &= \overline{\Im B} \\
\overline{H} &= \overline{S} + (\overline{\Im\tilde{A}} - \overline{S\tilde{C}})\overline{L} \\
\overline{\delta} &= \overline{L}
\end{aligned} \tag{5.12}$$

Equation (5.13) is obtained by subtracting Eq. (5.10) from Eq. (5.8), where $\dot{\overline{e}} = \tilde{\chi} - \hat{\tilde{\chi}}$,

$$\dot{\overline{e}} = (\overline{\Im\tilde{A}} - \overline{S\tilde{C}})\overline{e} \tag{5.13}$$

Based on **Assumption** 1, $(\overline{\Im\Xi}, \overline{\tilde{C}})$ is detectable, hence the matrix $(\overline{\Im\tilde{A}} - \overline{S\tilde{C}})$ is stable. Additionally, Eq. (5.13) indicates that the error dynamic asymptotically converges to zero if $(\overline{\Im\tilde{A}} - \overline{S\tilde{C}})$ is designed to be stable. The proof is completed.

Remark 1 The dynamic equation of the error in Eq. (5.13) indicates that the observer error is not affected by the faults or disturbances because disturbances (including system uncertainty and external disturbances) are considered to enhance the system state. In addition, the controller effort can be troubleshooted according to **Definition** 1; thus, in this case, the switch control input should not be considered.

Case *ii.* $0 < \overline{\Gamma}_i(.) < 1$

The following Assumption 2 is made to prove this case:

Assumption 2 The system fault $\overline{\Phi}$ and perturbations $\overline{\Psi}$ are norm-bounded; thus, $\|\overline{\Phi}\| \leq \overline{\rho}_1$ and $\|\overline{\Psi}\| \leq \overline{\rho}_2$, where $\overline{\rho}_1 > 0, \overline{\rho}_2 > 0$. In this case, the fault can be expressed as the sum of the matched and unmatched parts, as indicated in Eq. (5.14)

$$\overline{\alpha\Phi} = \overline{\alpha}_1\overline{\Phi}_1 + \overline{\alpha}_2\overline{\Phi}_2. \tag{5.14}$$

where $\overline{\Phi}_1$ is the matched part, $\overline{\alpha}_1 = \overline{a}_1\overline{B}$, and Φ_2 is the unmatched part. Equation (5.15) is obtained by substituting Eq. (5.14) into Eq. (5.2)

$$\overline{\Xi}\dot{\tilde{\overline{\chi}}} = \tilde{\overline{A}}\tilde{\overline{\chi}} + \overline{B}(u + \overline{a}_1\overline{\Phi}_1) + \overline{\alpha}_2\overline{\Phi}_2$$
$$\overline{\Upsilon} = \tilde{\overline{C}}\tilde{\overline{\chi}} \tag{5.15}$$

According to **Definition** 1, the matched part of the fault can be considered a part of a space spanned by the control input vector u. We multiply both sides of the first equation in Eq. (5.15) with the matrix $\overline{\Im} \in R^{(n+q)\times n}$ to obtain the following equation:

$$\overline{\Im\Xi}\dot{\tilde{\overline{\chi}}} = \overline{\Im\tilde{A}}\tilde{\overline{\chi}} + \overline{\Im B}[u + \overline{a}_1\overline{\Phi}_1] + \overline{\Im}\overline{\alpha}_2\overline{\Phi}_2. \tag{5.16}$$

Denote

$$u'' = [u + \overline{a}_1\overline{\Phi}_1]. \tag{5.17}$$

Then Eq. (5.16) can be rewritten as Eq. (5.18)

$$\overline{\Im\Xi}\dot{\tilde{\overline{\chi}}} = \overline{\Im\tilde{A}}\tilde{\overline{\chi}} + \overline{\Im B}u'' + \overline{\Im}\overline{\alpha}_2\overline{\Phi}_2. \tag{5.18}$$

By adding $\overline{L\Upsilon}$ to both sides of Eq. (5.18), the following equation can be obtained:

$$\dot{\tilde{\overline{\chi}}} = \overline{\Im\tilde{A}}\tilde{\overline{\chi}} + \overline{\Im B}u'' + \overline{L\Upsilon} + \overline{\Im}\overline{\alpha}_2\overline{\Phi}_2. \tag{5.19}$$

For the system with no mismatched fault parts (5.19), the virtual Luenberger observer (5.10) can be used, and the parameters of the observer (5.3) can be solved according to the method in (5.12). To compensate for $\overline{\Phi}_2$, the switch variables υ should be designed to eliminate the failures $\overline{\Phi}_2$. Therefore, the observer gain matrix $\overline{\Gamma\varpi}$ can be designed as $rank(\overline{\Gamma\varpi}, \overline{\alpha}_2) = rank(\overline{\Gamma\varpi})$. In addition, because the SMC is robust to faults and disturbances, it can be used to design discontinuous inputs υ to compensate $\overline{\Phi}_2$ for impacts. The sliding surface can be defined by Eq. (5.20):

$$\overline{s}(\overline{e}) = \overline{\lambda}[\overline{e} + \overline{\ell}sig^{\overline{a}}(\dot{\overline{e}})]. \tag{5.20}$$

where $sig^{\overline{a}}(\dot{\overline{e}}) = \left\|\dot{\overline{e}}\right\|^{\overline{a}}.sat(\dot{\overline{e}}), 0 < \overline{a} < 1, \overline{\lambda}$ and $\overline{\ell}$ are parameter matrices to be designed. $sat(.)$ is the saturation function. Based on the sliding surface (5.20), the input of the designed SMC can be written as follows:

$$\upsilon = -\overline{K}sign(\overline{s}) \tag{5.21}$$

where k is a scalar based on ρ_1 and ρ_2, which are predefined in **Assumption** 2. In addition, $sign(e)$ stands for a sign function and is defined as:

$$sign(\overline{e}) = \begin{cases} 1 & \overline{e} > 0 \\ 0 & \overline{e} = 0 \\ -1 & \overline{e} < 0 \end{cases}.$$

The proof is completed.

Remark 2 In this case, the switching input variable υ in observer (5.3) will partially work according to the value of $\overline{\Gamma}(.)$ to compensate only for $\overline{\Phi}_2$, since for the fault part, $\overline{\Phi}_1$ will be compensated for as in the previous **case i** by the control effort u.

Case *iii*. $\overline{\Gamma}_i(.) = 1$

When the value of k is set to ∞ in Eq. (5.4), $\overline{\Gamma}_i(.)$ becomes a step function, and the incipient fault becomes an abrupt fault. In this case, the augmented system and the ASMMSO observer take the forms of Eqs. (5.22) and (5.23), respectively.

$$\begin{aligned} \overline{\Xi}\dot{\tilde{\overline{\chi}}} &= \overline{\tilde{A}}\tilde{\overline{\chi}} + \overline{B}u + \overline{\alpha}\overline{\Phi} \\ \overline{\Upsilon} &= \overline{\tilde{C}}\tilde{\overline{\chi}} \end{aligned} \tag{5.22}$$

$$\begin{aligned} \dot{\overline{\Omega}} &= \overline{\mathrm{M}\Omega} + \overline{\varphi}u + \overline{\mathrm{H}\Upsilon} + \overline{\varpi}\upsilon \\ \hat{\tilde{\overline{\chi}}} &= \overline{\Omega} + \overline{\delta\Upsilon} \\ \hat{\overline{\Upsilon}} &= \overline{\tilde{C}}\hat{\tilde{\overline{\chi}}} \end{aligned} \tag{5.23}$$

We left-multiply both sides in the first equation of the augmented system (5.22) by $\overline{\Im} \in R^{(n+q)\times n}$ and denote a new variable $\overline{\gamma}$ as follows:

$$\overline{\gamma} = \overline{\Im\Xi}\tilde{\overline{\chi}} - \overline{\Omega} \tag{5.24}$$

Based on Eqs. (5.22), (5.23), and (5.24), the following equation can be obtained.

$$\dot{\overline{\gamma}} = \overline{M\gamma} + (\overline{\Im\tilde{A}} - \overline{M\Im\Xi} + \overline{\mathrm{H}\tilde{C}})\tilde{\overline{\chi}} + (\overline{\Im B} - \overline{\varphi})u + \overline{\Im\alpha\Phi} - \overline{\varpi}\upsilon \tag{5.25}$$

Let the estimation error, $\overline{e} = \tilde{\overline{\chi}} - \hat{\tilde{\overline{\chi}}}$; then it can be expressed as $\overline{e} = \overline{\gamma} + (\mathrm{I}_{n+q} - \overline{\delta\tilde{C}} - \overline{\Im\Xi})\tilde{\overline{\chi}}$. We design the matrix equations as:

$$\begin{aligned} &\overline{\Im\tilde{A}} - \overline{M\Im\Xi} + \overline{\mathrm{H}\tilde{C}} = 0 \\ &\overline{\Im B} - \overline{\varphi} = 0 \\ &\mathrm{I}_{n+q} - \overline{\delta\tilde{C}} - \overline{\Im\Xi} = 0 \end{aligned} \tag{5.26}$$

After satisfying Eq. (5.26), the error dynamic can be written in the form of Eq. (5.27):

$$\dot{\overline{e}} = \overline{M}\overline{e} + \overline{\Im\alpha}\overline{\Phi} - \overline{\varpi}\upsilon \tag{5.27}$$

We design the following:

$$\overline{\Im\alpha} = \overline{E}^{-1}\overline{\tilde{C}}^{T}\overline{\vartheta} \tag{5.28}$$

$$\overline{\varpi} = \overline{E}^{-1}\overline{\tilde{C}}^{T} \tag{5.29}$$

The switching variable is designed as $\upsilon = \overline{\tau}\sin(\tilde{\overline{e}})$ to maintain the system in the sliding phase, where $\tilde{\overline{e}} = \overline{\Upsilon} - \hat{\overline{\Upsilon}}$, and $\overline{\tau}$ is a scalar. The discontinuous control is designed as follows:

$$\upsilon_{eq} = \frac{1}{T\overline{s}+1}\overline{\upsilon} \tag{5.30}$$

where T is the time constant of the filter. We substitute Eqs. (5.28), (5.29), and (5.30) into Eqs. (5.27) to obtain Eq. (5.31).

$$\dot{\overline{e}} = \overline{M}\overline{e} + \overline{E}^{-1}\overline{\tilde{C}}^{T}(\overline{\vartheta\Phi} - \upsilon_{eq}). \tag{5.31}$$

When the sliding motion occurs, the dynamic equation of the error becomes

$$\dot{\overline{e}} = \overline{M}\overline{e} \tag{5.32}$$

According to Eq. (5.32), if the matrix $\overline{M}$ is a Hurwitz matrix, the error converges to zero asymptotically.

Note: Since the stability analysis has been described in [15], we do not provide any details.

Remark 3 Compared to the FE/FTC methodology proposed in this study, the most recent schemes described in the literature (e.g. [14, 15]) are considered special cases since these methods are designed based on the worst-case scenario, and the fault evolution profile performance is sacrificed to achieve better robustness.

5.3.2 Design of Controller

In the next parts, we implement an STW-based SMC to construct the input control signal $u(t)$ to compensate for actuator failures and sensor interference. The controller design includes the following three stages: functional design of the sliding mode surface, extraction of estimation information, and construction of control signals.

Stage 1: Sliding mode surface function design.

The sliding mode surface function is designed as:

$$\overline{s}(t) = [\partial/\partial t + \overline{r}]^{n-1} \cdot \overline{\varsigma} \tag{5.33}$$

where $\overline{r} > 0$ is a scalar, and $\partial/\partial t$ represents the differential operator.

Note: In Eq. (5.33) n denotes the system order. For example, in a second-order system, the value of$n$ is 2; thus, the sliding surface Eq. (5.33) can be expressed as $\overline{s}(t) = \dot{\overline{\varsigma}} + k_1\overline{\varsigma}$, where $\overline{s}(t) = \ddot{\overline{\varsigma}} + k_2\dot{\overline{e\varsigma}} + k_1\overline{\varsigma}$ for $n = 3$, and so on.

Remark 4 In the siding- ode surface (5.33), when the sliding surface is reached, i.e. $\overline{s}(t) = 0$, the error variable $\overline{\varsigma}$ becomes an attractor.

Stage 2: Extracting the estimated information.

In this part, the FHOSM controller is implemented according to the state estimation $\hat{\overline{x}}$ and the disturbance $\hat{\overline{\Psi}}$ estimation information provided by the multistage observer (5.3). According to the second relation formula in the previous observer (5.3), the following relation can be obtained:

$$\begin{aligned}
\dot{\hat{\overline{\chi}}} &= \dot{\overline{\Omega}} + \dot{\overline{\delta\Upsilon}} \\
&= \overline{M\Omega} + \overline{\varphi}u + \overline{H\Upsilon} + \overline{\Gamma}\overline{\varpi}\upsilon + \dot{\overline{\delta\Upsilon}} \\
&= \overline{M}(\hat{\overline{\chi}} - \overline{\delta\Upsilon}) + \overline{H\Upsilon} + \overline{\varphi}u + \overline{\Gamma}\overline{\varpi}\upsilon + \dot{\overline{\delta\Upsilon}} \\
&= \overline{M}\hat{\overline{\chi}} + (\overline{H} - \overline{M\delta})\overline{\Upsilon} + \overline{\varphi}u + \overline{\Gamma}\overline{\varpi}\upsilon + \dot{\overline{\delta\Upsilon}} \\
&= [\overline{M} + (\overline{H} - \overline{MH})\tilde{\overline{C}}]\hat{\overline{\chi}} + \overline{\varphi}u(k) + \overline{\Gamma}\overline{\varpi}\upsilon + \overline{\delta}\tilde{\overline{C}}\dot{\hat{\overline{\chi}}}
\end{aligned} \tag{5.34}$$

By rearranging Eq. (5.34) and solving it for $\dot{\hat{\overline{\chi}}}$, Eq. (5.35) is obtained:

$$\dot{\hat{\overline{\chi}}} = (I - \overline{\delta}\tilde{\overline{C}})^{-1}\{[\overline{M} + (\overline{H} - \overline{M\delta})\tilde{\overline{C}}]\hat{\overline{\chi}} + \overline{\varphi}u(k) + \overline{\Gamma}\overline{\varpi}\upsilon\} \tag{5.35}$$

The augmented system state $\hat{\overline{\chi}}$ in Eq. (5.2) should be filtered to obtain the estimation information on the system states and system perturbations.

We use $\overline{\mu}_1 = (I - \overline{\delta}\tilde{\overline{C}})^{-1}[\overline{M} + (\overline{H} - \overline{M\delta})\tilde{\overline{C}}]$,$\overline{\mu}_2 = (I - \overline{\delta}\tilde{\overline{C}})^{-1}\overline{\varphi}$, and $\overline{\mu}_3 = (I - \overline{\delta}\tilde{\overline{C}})^{-1}\overline{\Gamma}\overline{\varpi}$. From the model Eq. (5.1), it can be concluded that $\hat{\overline{\chi}} = \left[\hat{\overline{x}}\ \hat{\overline{\Psi}}\right]^T$; thus:

$$\begin{pmatrix} \dot{\hat{\overline{x}}} \\ \dot{\hat{\overline{\Psi}}} \end{pmatrix} = \begin{pmatrix} \overline{\mu}_{11} & \overline{\mu}_{12} \\ \overline{\mu}_{13} & \overline{\mu}_{14} \end{pmatrix}\begin{pmatrix} \hat{\overline{x}} \\ \hat{\overline{\Psi}} \end{pmatrix} + \begin{pmatrix} \overline{\mu}_{21} \\ \overline{\mu}_{22} \end{pmatrix}u + \begin{pmatrix} \overline{\mu}_{31} \\ \overline{\mu}_{32} \end{pmatrix}\upsilon \tag{5.36}$$

According to Eq. (5.36), the following relationship can be obtained:

$$\dot{\hat{\overline{x}}} = \overline{\mu}_{11}\hat{\overline{x}} + \overline{\mu}_{21}u + \overline{\mu}_{31}\upsilon + \overline{\mu}_{12}\hat{\overline{\psi}} \tag{5.37}$$

$$\dot{\hat{\overline{\Psi}}} = \overline{\mu}_{13}\hat{\overline{x}} + \overline{\mu}_{14}\hat{\overline{\psi}} + \overline{\mu}_{22}u + \overline{\mu}_{32}\upsilon \tag{5.38}$$

Stage 3: Constructing the control signal.

Assumption 3 In the model (5.1), the pair $(\overline{A}, \overline{B})$ is controllable, and rank $\overline{\beta}_s = \overline{q}$.

According to **Assumption** 3 and using the appropriate linear transformation $\overline{\chi} = \overline{P}_c\overline{x}$ (where $\overline{P}_c$ is the non-singular transformation matrix), the system (5.1) can be transformed into a controllable canonical form, such as:

$$\begin{bmatrix} \overline{x}_1 \\ \vdots \\ \overline{x}_n \end{bmatrix} = \begin{bmatrix} 0 & 1 & 0 & \cdots & 0 \\ 0 & 0 & 1 & \cdots & 0 \\ \vdots & \vdots & \vdots & \ddots & \vdots \\ 0 & 0 & 0 & \cdots & 1 \\ -\overline{a}_n & -\overline{a}_{n-1} & -\overline{a}_{n-2} & \cdots & -\overline{a}_1 \end{bmatrix} \begin{bmatrix} \overline{x}_1 \\ \vdots \\ \overline{x}_n \end{bmatrix} + \begin{bmatrix} 0 \\ \vdots \\ 0 \\ 1 \end{bmatrix} u + \begin{bmatrix} \overline{\beta}_{\overline{a}1} \\ \vdots \\ \overline{\beta}_{\overline{a}n} \end{bmatrix} \begin{bmatrix} \overline{\Psi}_1 \\ \vdots \\ \overline{\Psi}_n \end{bmatrix} \tag{5.39}$$

Let the tracking error be:

$$\overline{\varsigma} = \overline{x}_1 - \overline{x}_{1d} \tag{5.40}$$

where $\overline{x}_1$ is the system output, which is required to track the reference signal $\overline{x}_{1d}$ that can be expressed as a step function. Its derivative $\dot{\overline{x}}_{1d} = 0$.

Based on Eqs. (5.33), (5.39), and the tracking error (5.40), the derivative of the sliding surface is:

$$\begin{aligned} \dot{\overline{s}}(t) &= k_1\overline{x}_2 + k_2\overline{x}_3 + \cdots + k_{n-2}\overline{x}_{n-1} - [\overline{a}_n\overline{x}_1 + \overline{a}_{n-1}\overline{x}_2 \\ &\quad + \cdots + \overline{a}_1\overline{x}_n + u] + \tilde{\overline{\beta}}_a\tilde{\overline{\Psi}} \\ &= -\overline{a}_n\overline{x}_1 - \overline{a}_1\overline{x}_n + (k_1 - \overline{a}_{n-1})\overline{x}_2 + (k_2 - \overline{a}_{n-2})\overline{x}_3 \\ &\quad + \cdots + (k_{n-2} - \overline{a}_2)\overline{x}_{n-1} + u + \tilde{\overline{\beta}}_a\tilde{\overline{\Psi}} \end{aligned} \tag{5.41}$$

where $\tilde{\overline{\beta}}_a\tilde{\overline{\Psi}}$ represents the lumped perturbations.

In this paper, the STW algorithm (5.42) is employed to design the control variable u to eliminate the system perturbations and the chattering.

$$\begin{aligned} u &= \overline{\varepsilon}_1||\overline{s}||^{1/2}sign(\overline{s}) - X \\ X &= -\overline{\varepsilon}_2 sign(\overline{s}) \end{aligned} \tag{5.42}$$

where $\overline{\varepsilon}_1, \overline{\varepsilon}_1$ are designed as in [24], and $\hat{\overline{\Psi}}$ is the disturbance estimation. By substituting Eq. (5.42) into Eq. (5.41), we obtain the equivalent and reaching control law in Eqs. (5.43) and (5.44), respectively:

$$\begin{aligned} u_{eq} &= \overline{a}_n\overline{x}_1 + \overline{a}_1\overline{x}_n - (k_1 - \overline{a}_{n-1})\overline{x}_2 - (k_2 - \overline{a}_{n-2})\overline{x}_3 \\ &\quad - \cdots - (k_{n-2} - \overline{a}_2)\overline{x}_{n-1} - \overline{\varepsilon}_1||\overline{s}||^{1/2} sign(\overline{s}) - \int \overline{\varepsilon}_2 sign(\overline{s})) \end{aligned} \tag{5.43}$$

$$u_{re} = -\tilde{\overline{\beta}}_a\tilde{\overline{\Psi}} \tag{5.44}$$

Finally, by replacing $\overline{x}$ and $\tilde{\overline{\Psi}}$ with $\hat{\overline{x}}$ and $\hat{\overline{\Psi}}$ in Eqs. (5.43) and (5.44), the FHOSM controller can be obtained as:

$$u = u_{eq} + u_{re}. \tag{5.45}$$

5.3.3 Analysis of System Behavior

Definition 2 For continuous-time systems, sliding motion can be achieved if the derivative of the candidate positive definite Lyapunov function is negative definite. Furthermore, the upper limit of the time of arrival depends on the steepest descent slope of the designed sliding surface.

Theorem 2 *We use the FHOSM controllers in sliding surfaces* (5.33) *and* (5.45) *for the closed-loop perturbation system model defined in Eq.* (5.47)*. If Assumptions 3 and 4 are satisfied, all trajectories of the sliding surfaces starting from any position will reach the side-mode manifold in a finite time, and the time derivative of the Lyapunov function* (5.48) *is negative.*

Proof Substituting (5.45) into (5.41) yields:

$$\dot{\overline{s}} = -\overline{\varepsilon}_1 \| \overline{s} \|^{1/2} sign(\overline{s}) - \int \overline{\varepsilon}_2 sign(\overline{s}) + \overline{\lambda} \tag{5.46}$$

Remark 5 Although **Theorem** 1 proves that the estimation error is bounded and converted to zero, the transient behavior of the closed-loop FTC system is affected. Therefore, during the controller design, a new variable $\overline{\lambda}$is introduced to compensate automatically for the impact of the estimation error.

We define $\overline{s} = \overline{\zeta}_1$ and $-\int \overline{\varepsilon}_2 sign(\overline{s}) = \overline{\zeta}_2$, and Eq. (5.46) can be rewritten as:

$$\begin{aligned} \dot{\overline{\zeta}}_1 &= -\overline{\varepsilon}_1 \| \overline{\zeta}_1 \|^{1/2} sign(\zeta_1) + \zeta_2 + \overline{\lambda} \\ \dot{\overline{\zeta}}_2 &= -\overline{\varepsilon}_2 sign(\zeta_1) \end{aligned} \tag{5.47}$$

Assumption 4: The effect of the estimation error $\tilde{\lambda}$ is bounded, and a known positive number $\overline{\hbar}$ can be found to satisfy $\left\|\tilde{\lambda}\right\| \leq \overline{\hbar}\left\|\overline{\zeta}_1\right\|^{\frac{1}{2}}$.

The present system (5.47) is a general STW, as demonstrated in [24]. Therefore, the candidate Lyapunov functions may be similar. In this chapter, we define the Lyapunov function of system (5.47) as:

$$V(\overline{\zeta}_1) = \overline{Z}^T \overline{N Z} \tag{5.48}$$

where $\overline{N} = \frac{1}{2}\begin{bmatrix} 4\overline{\varepsilon}_2 + \overline{\varepsilon}_1^2 & -\overline{\varepsilon}_1 \\ -\overline{\varepsilon}_1 & 2 \end{bmatrix}$, and $\overline{Z}^T = \begin{bmatrix} \left\|\overline{\zeta}_1\right\|^2 sign(\overline{\zeta}_1) & \overline{\zeta}_2 \end{bmatrix}$.

According to **Definition** 2, the derivative of the Lyapunov function (5.48) should be negative to confirm that the closed-loop system dynamics (5.47) will converge to zero so that:

$$\dot{V}(\overline{\zeta}_1) = \frac{-1}{\left\|\overline{\zeta}_1\right\|^{1/2}} \overline{Z}^T X \overline{Z} + \frac{\tilde{\lambda}}{\left\|\overline{\zeta}_1\right\|^{1/2}} \overline{\ell}^T \overline{Z} \tag{5.49}$$

where:

$X = \frac{\overline{\varepsilon}_1}{2}\begin{bmatrix} 2\overline{\varepsilon}_2 + \overline{\varepsilon}_1^2 & -\overline{\varepsilon}_1 \\ -\overline{\varepsilon}_1 & 1 \end{bmatrix}$ And $\overline{\ell}^T = \begin{bmatrix} 2\overline{\varepsilon}_2 + \frac{\overline{\varepsilon}_1^2}{2} & -\frac{\overline{\varepsilon}_1}{2} \end{bmatrix}$

If **Assumption 4** is satisfied, then:

$$\dot{V}(\overline{\zeta}_1) \leq -\frac{1}{\left\|\overline{\zeta}_1\right\|^{1/2}} \overline{Z}^T \overline{W Z} \tag{5.50}$$

where

$$\overline{W} = \frac{\overline{\varepsilon}_1}{2}\begin{bmatrix} 2\overline{\varepsilon}_2 + \overline{\varepsilon}_1^2 - 2\overline{\hbar} & * \\ -\overline{\varepsilon}_1 + \frac{2\overline{\hbar}}{\overline{\varepsilon}_1} & 1 \end{bmatrix} \tag{5.51}$$

We design $\begin{cases} \overline{\varepsilon}_1 > 2\overline{\hbar} \\ \overline{\varepsilon}_2 > \overline{\varepsilon}_1 \frac{5\overline{\hbar}\overline{\varepsilon}_1 + 4\overline{\hbar}^2}{2(\overline{\varepsilon}_1 - 2\overline{\hbar})} \end{cases}$ so $\overline{W} > 0$ which implies that the derivative of the candidate Lyapunov function is negative definite. This finishes the proof.

5.4 Simulation Results

In the simulation, the parameters of the networked DC motor system (5.1) are defined as follows:

$\overline{A} = \begin{bmatrix} -2.219 & 4.438 \\ -12.000 & -24.000 \end{bmatrix}$, $\overline{B} = \begin{bmatrix} 0 \\ 20 \end{bmatrix}$, $\overline{C} = \begin{bmatrix} 10 \\ 01 \end{bmatrix}$, $\overline{\alpha} = \begin{bmatrix} 0.001 \\ 19.999 \end{bmatrix}$, $\overline{\beta}_a = \begin{bmatrix} 0.1000 \\ 0.1000 \end{bmatrix}$ and $\overline{\beta}_s = \begin{bmatrix} 1 \\ 0 \end{bmatrix}$.

The observer parameter matrices in cases i and ii are:

$$\overline{M} = \begin{bmatrix} -3.6633\ 3.5496 - 0.8963 \\ -6.4111\ \text{-}\ 13.82270.0534 \\ 1.7734\ \text{-}\ 5.3285 - 0.0963 \end{bmatrix}, \overline{\varphi} = \begin{bmatrix} 0.7406 \\ 10.6850 \\ 0.7425 \end{bmatrix}, \overline{H} = \begin{bmatrix} 0.1037 & 3.1067 \\ 0.0534 & -5.2902 \\ -0.0963 & -4.0504 \end{bmatrix}$$

$$\overline{\delta} = \begin{bmatrix} 0.8960 & -26.5286 \\ -1.4081 & 53.9909 \\ -0.0914 & 34.0882 \end{bmatrix}, \overline{\varpi} = \begin{bmatrix} 0.0378 & 0 \\ 0 & 0.0378 \\ 0.0378 & 0 \end{bmatrix}.$$

The observer parameter matrices for the third case in (5.26) are the same as in [15]. The controller parameters are $\overline{\varepsilon}_1 = 11$ and $\overline{\varepsilon}_2 = 12$. In addition, the reference angular speed velocity is one.

The disturbances and faults of the systems are set as in [15]:

$$\overline{\Psi}(t) = \begin{cases} 0.05\sin(\pi t), & 0 \le t \le 10s \\ 3\sin(4\pi t) + [0.10.5]x, & 10s < t < 15s \end{cases}$$

$$\overline{\Phi}(t) = \begin{cases} 0, & 0s \le t \le 2s \\ 0.04(t-2)^2 + \sin(\pi(t-2)), & 2s < t < 7s \\ 1, & 7s < t < 10s \\ 2\sin(3\pi(t-10)) + 1, & 10s < t \le 15s \end{cases}$$

The matching time $T_C \ge 7$ and the value of k are designed as follows:

$$k = \begin{cases} 1 < k \le 2 & 7s \le T_c \le 10s \\ 200 < k \le \infty & 10s < T_c \le 15s \end{cases}$$

The simulation results are shown as follows.

Figure 5.3 contains four different zones. The first represents the normal operating zone. The backstepping controller is unable to maintain the motor's angular velocity at the expected 1 rad/Sec in this region. In the second zone, faults begin to occur. The backstepping controller oscillates between 0.8 and 1 rad/Sec, while the proposed FHOSM controller keeps the motor speed at 1 rad/Sec. In the third zone, the failure becomes an unbeatable type of failure with a very low rating, demonstrating the robustness of the FHOSM controller. The last area exhibits a high level of mismatched faults and high measurement noise. Although both controllers oscillate during this period, the oscillation level of the FHOSM controller is much smaller than that of the reverse controller.

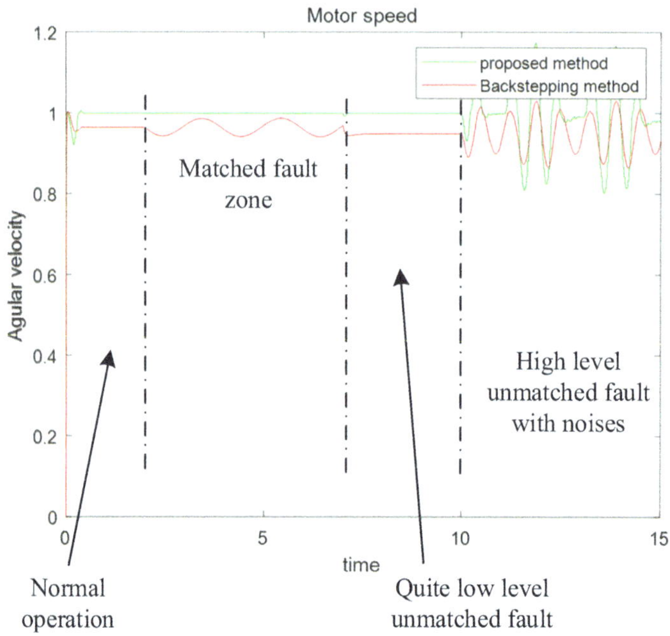

Fig. 5.3 Angular velocity for differentiable fault types

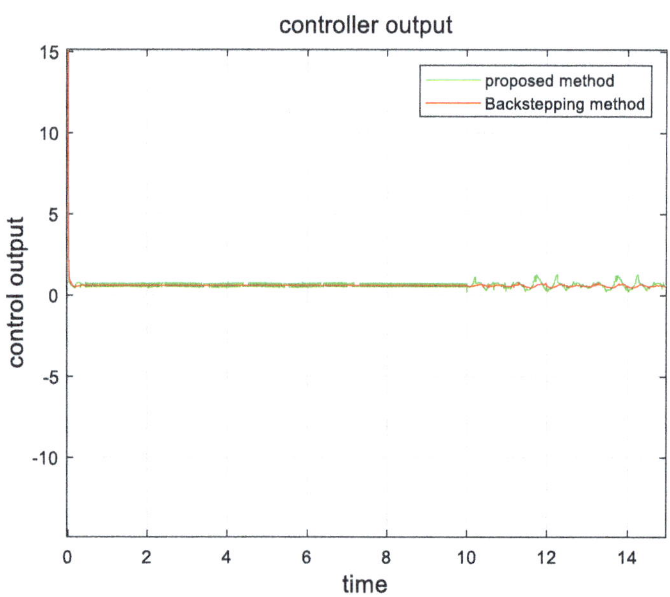

Fig. 5.4 Controller output for differentiable fault type

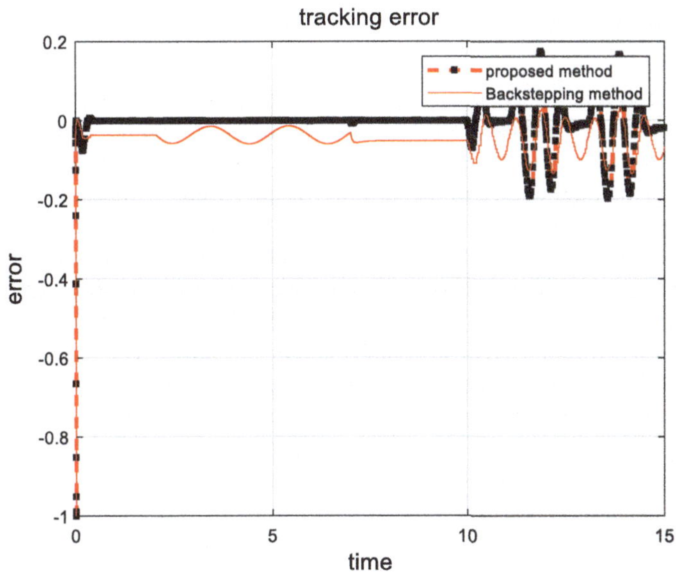

Fig. 5.5 Tracking error for differentiable fault type

Figure 5.4 shows the output actions of the proposed FHOSM controller and back-stepping controller. In addition, the figure shows the sliding action of the proposed controller, demonstrating why the performance of the proposed controller is more robust than that of the pushback controller.

In Fig. 5.5, the backstepping controller oscillates at the beginning of the fault period, whereas the FHOSM controller retains robust properties throughout all periods, except for very low levels of oscillation and sensor measurement noise in the mismatched fault period.

Therefore, the proposed control method is not only designed for the worst case but also preserves the nominal operating state and low-level fault condition to reduce energy consumption. It is also worth noting that, in the absence of faults, the proposed method is compact and represents an interference-observer-based control technique that has attracted great attention in many industrial applications. The limitation of this study is that the control system cannot automatically detect the fault level. In addition, the proposed controllers focus on SISO linear systems with disturbances and faults; however, in practice, most applications are multiple-input multiple-output (MIMO) nonlinear systems. These limitations will be considered in a future study.

5.5 Conclusions

In this chapter, a new FTC method was proposed to deal with actuator and sensor fault/disturbance problems in NCSs. First, an adaptive sliding mode multi-stage observer was designed to model the fault change over time and compensate for faults according to the fault degree of the actuator. Then, based on the estimated state, an FHOSM controller was designed to achieve finite-time convergence without jitter. The duration sliding function was used. In addition, the system dynamics under the proposed control were analyzed, and the superiority of the proposed algorithm was demonstrated using the Lyapunov method. Finally, the proposed algorithm was tested using a DC motor NCS with faults and disturbances. The simulation results showed that the proposed controller could adequately protect the NCS from faults and interference in the actuator and sensor channels.

References

1. M. Li and Y. Chen, "Challenging research for networked control systems-a survey," *Transactions of the Institute of Measurement and Control*, vol. 41, no. 9, pp. 2400-2418, 2019.
2. Amin, A. Ahmed, Hasan, and K. Mahmood, "A review of fault tolerant control systems: advancements and applications," Measurement, vol.143, pp.58–68, 2019.
3. D. Ye, M. Chen, and K. Li, "Observer-based distributed adaptive fault-tolerant containment control of multi-agent systems with general linear dynamics" ISA Transactions, vol.71, pp.32-39, 2017.
4. Y. Liu, G. Yang, and X. Li, "Fault-tolerant control for uncertain linear systems via adaptive and LMI approaches," International Journal of Systems Science, vol.48, no. 2, pp.347-356, 2017.
5. M. Li and Y Chen, "Robust tracking control of networked control systems with communication constraints and external disturbance," IEEE Transactions on Industrial Electronics, vol. 64, no. 5, pp. 4037-4047, 2017.
6. M. Blanke, M. Kinnaert, J. Lunze, and M. Staroswiecki, "Diagnosis and fault-tolerant control," Springer Berlin Heidelberg, 2006.
7. W. Chengwei, L. Jianxing, X. Yongyang, and W. Ligang, "Observer-based adaptive fault-tolerant tracking control of nonlinear Nonstrict-Feedback Systems," IEEE Transactions on Neural Networks and Learning Systems. vol. 29, no. 7, pp.3022-3033, 2018.
8. J. Wang and G. Yang, "Data-driven output-feedback fault-tolerant tracking control Method and its application to a DC Servo System," IEEE/ASME Transactions on Mechatronics, vol.24, no. 3, 2019.
9. L. Hao and J. Park, "Integral sliding mode fault-tolerant control for uncertain linear systems over networks with signals quantization," IEEE Transactions on Neural Networks and Learning Systems, vol. 28, 2017.
10. X. Liu, Z. Gao, and A. Zhang, "Robust fault tolerant control for discrete-time dynamic systems with applications to aero engineering systems," IEEE Transactions Access, vol. 6, 2018.
11. J. Lan and R. Patton, "A new strategy for integration of fault estimation within fault-tolerant control." Automatica, vol. 69, pp. 48–59, 2016.
12. A. Loza, J. Cieslak, D. Henry, A. Zolghadri, and L. Fridman, "Output tracking of systems subjected to perturbations and a class of actuator faults based on HOSM observation and identification," Automatica, vol. 59, pp.200–205, 2015.

13. A. Teixeira, J. Araújo, H. Sandberg, and K. Johansson, "Distributed sensor and actuator reconfiguration for fault-tolerant networked control systems" IEEE Transactions on Control of Network Systems, vol. 5, no. 4, pp.1517-1528, 2018.
14. M. Li and Y. Chen, "Robust adaptive sliding mode control for switched networked control systems with disturbance and faults," IEEE Transactions on Industrial Informatics, vol. 15, no. 1, pp. 193 - 204, Jan. 2019.
15. J. Lan and R. Patton, "A decoupling approach to integrated fault-tolerant control for linear systems with unmatched non-differentiable faults," Automatica, Vol.89, pp.290-299, 2018.
16. M. Corradini and A. Cristofaro, "Fault-tolerant control for linear systems with multiple faults and disturbances based on augmented intermediate estimator," IET Control Theory & Applications, vol. 11, Iss. 2, pp. 164–172, 2017.
17. Y. Chen and B. Guo, "Sliding mode fault tolerant tracking control for a single-link flexible joint manipulator system" IEEE Access, vol. 31, July 10, 2019.
18. R. Guzman, L.G. Vicu˜na, J. Morales, M. Castilla, and J. Matas, "Sliding-mode control for a three-phase unity power factor rectifier operating at fixed switching frequency," IEEE Transactions on Power Electronics, vol. 31, no. 01, pp. 758–769, 2016.
19. Q. Xu, "Piezoelectric nanopositioning control using second-order discrete-time terminal sliding-mode strategy," IEEE Transactions on Industrial Electronics, vol. 62, no. 12, pp. 7738-7748, Dec. 2015.
20. S. Li, C. Wu, and Z. Sun, "Design and implementation of clutch control for automotive transmissions using terminal-sliding-mode control and uncertainty observer," IEEE Transactions on Vehicular Technology, vol. 65, no. 4, pp. 1890-1898, 2016.
21. J. Davila, L. Fridman, and A. Levant, "Second-order sliding-mode observer for mechanical systems," IEEE Transactions on Industrial Informatics, vol. 15, no. 1, pp. 193 - 204, 2019.
22. M. Van, H.-J. Kang, Y.-S. Suh, and K. S. Shin, "Second order sliding mode-based output feedback tracking control for uncertain robot manipulators" International Journal of Advanced Robotic Systems., vol. 10, no. 16, pp. 1–9, 2013.
23. M. Van, H. Kang, and K. Shin, "Novel quasi-continuous super-twisting high-order sliding mode controllers for output feedback tracking control of robot manipulators," Journal of Mechanical Engineering Science., vol. 228, no. 17, pp.3240-3257, 2014.
24. Khalil and M. Osorio, "A Lyapunov approach to second-order sliding mode controllers and observers," in Proc. 47th IEEE Conf. Decis. Control, Cancún. Mexico, pp. 2856–2861, 2008.
25. W. Gao, Y. Wang, and A. Homaifa, "Discrete-time variable structure control systems," IEEE Transactions on Industrial Electronics, vol. 42, no. 2, pp. 117-122, Apr. 1995.
26. S. Qu, X. Xia, and J. Zhang, "Dynamics of discrete-time sliding-mode-control uncertain systems with a disturbance compensator," IEEE Transactions on Industrial Electronics, vol. 61, no. 7, pp. 3502- 3510, Jul. 2014.
27. M. Van, S. Ge, and H. Ren, "Finite time fault tolerant control for robot manipulators using time delay estimation and continuous nonsingular Fast Terminal Sliding Mode Control," IEEE Transactions on Cybernetics, vol. 47, no. 7, pp. 1681 - 1693, Jul. 2017.
28. H. Ma, J. Wu, and Z. Xiong, "Discrete-time sliding-mode control with improved quasi-sliding-mode domain," IEEE Transactions on Industrial Electronics, vol. 63, no. 10, pp. 6292-6304, Oct. 2016.

Chapter 6
Finite-Time Sliding-Mode Tracking Control for Networked Systems with Faults and Disturbances

In this chapter, the fault and interference problems in networked control systems (NCS) with discontinuous communication (time delay, data loss, and data chaos) are investigated. A finite-time controller based on a sliding mode algorithm is designed to compensate for system faults and eliminate interference affecting actuators and sensors. First, the well-known extended state observer is used to evaluate the system state and compensate for actuator failures simultaneously. Then, a new sliding surface function is established based on the non-saturated fast descending terminal sliding mode scheme (NFDTSMC), in which a fast terminal sliding mode controller (FTSMC) is obtained based on the online estimation information. In addition, data filters are designed, and a new error stamp order (ESO) is introduced to address network defects. In addition, the proposed control law is used to analyze the system performance, and the stability of the proposed method is proved. Finally, in the presence of network constraints, the simulation results show the advantages of this method for compensating for system faults and interference and suppressing network problems.

6.1 Introduction

The concept of an NCS means that the system contains intelligent sensors, controllers, and actuators coupled through a wireless communication network. In recent years, due to the emergence of technical tools and Internet facilities, the so-called NCS has gradually appeared in control systems [1]. The advantages of the NCS over traditional control systems include source allocation, low price, ease of expansion, moderate fixing and maintenance costs, as well as extraordinary robustness [2, 3]. Therefore, the NCS permeates most industrial and application fields, such as remote transportation systems and cloud control systems [4–6]. However, the NCS also introduces many novel complexities, such as network challenges, faults, interference, time delays, packaging faults, and flutter issues [7–9]. Among these limitations,

© The Author(s), under exclusive license to Springer Nature Singapore Pte Ltd. 2022

M. Li et al., *Tracking Control of Networked Systems via Sliding-Mode*,
https://doi.org/10.1007/978-981-16-6514-1_6

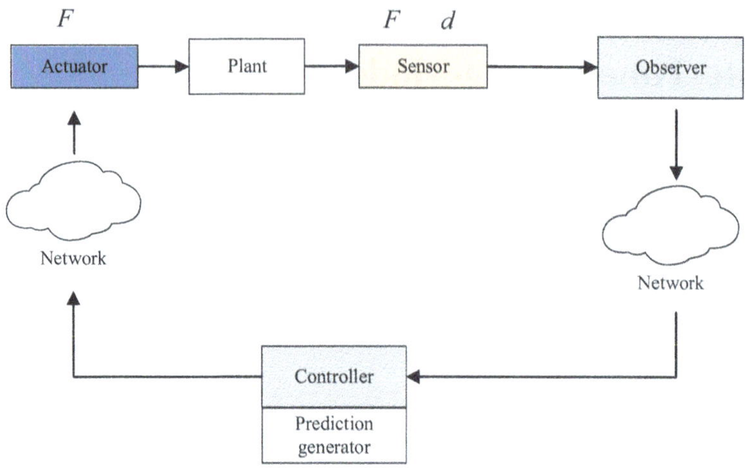

Fig. 6.1 Structure of NCS system with faults and disturbances

system failures, interference time delays, and packers are the most common problems that can compromise system quality and affect system stability. The configuration of an NCS with actuator failure and sensor interference is shown in Fig. 6.1, where the sensors and actuators are located on the factory side.

The main function of the sensor is to monitor the variable parameters of the system, such as the temperature, potential, and angular velocity, and provide the measurement results to the controller through the wired/wireless network feedback channel. The controller processes the input information and sends the control signal to the actuator through the NCS channel. After the actuator successfully obtains the controller signal, it can drive the system. At the same time, failures F can erode the actuator and create interference d in the actuator and sensors. As a result, the NCS will lose its ability to link the system components, and the system will be completely corrupted. Several approaches have been introduced to solve NCS problems and address the above challenges. For example, the network predictive control scheme (NPCS) was proposed to mitigate inevitable and common network constraints, such as time delays and packet loss in network communication [10–15]. The basic goal of the controller design scheme is to predict the system's future performance. At any given time, control measures are based on online monitoring data of packet intervals, failures, and packet failures. Liu et al. proposed an important NPC algorithm in [16]. The technology is classified as a controller based on predictive models. In addition, it retains the important attributes of an NCS. In [17], an enhanced NPC method was designed to compensate for network time delay and signal loss. One advantage of this approach is that it can be applied to several different industries. Sun et al. [18] searched for problems occurring in network feedback and forwarding channels, such as communication delays and unsuccessful packet arrivals. In this work, a new controller scheme is proposed to compensate for the influence of network defects. However, in practical applications, the controlled system will inevitably be subject to substantial external interference [19]. From another perspective, any part of any NCS can fail. For example, sensor and actuator failures are the most common types of failure. Because of their destructive effect on the control system, they may destabilize the closed-loop system [20]. According to the author's knowledge, the previously mentioned problems may reduce the performance of the control algorithm. Moreover, these limits challenge the control designer's barriers and pose economic challenges. However, for random interference in NCSs, faults and communication limitations are still in the infancy stage of research. This paper aims to address previous research gaps.

Sliding mode control (SMC) is an intelligent control method with a specific and variable structure. Utkin first proposed this method in 1977 [21]. The nonlinear control action enables the SMC to provide robust control characteristics that are independent of the system under different types of faults and external disturbances such as sensors or actuators fault and communication restrictions [22, 23]. Despite the advantages of SMC, it has an inherent jitter problem that affects its applications. Several studies proposed solutions to the jitter problem in recent years, such as the higher-level sliding mode [24] and a controller-based adaptive scheme [25]. In [26], Gao et al., for the first time, designed a new sliding mode-based controller to alleviate the oscillation problem [25]. In [27], a robust finite-time-based controller was designed based on fault-tolerant control, and a new non-singular fast terminal attractor technique (NFTAT) was used for passive and active control (PFTC/AFTC) to compensate for uncertain systems. In [28], a digital controller is adopted by using the velocity evaluation structure of the discrete-time sliding mode controller.

Based on the above SMC literature, the faults, disturbances (including the uncertainty of the controlled system), and network communication constraints in an NCS are studied. An FTSMC is proposed to prevent system failure and interference. The first design step is to construct an extended observer to generate the system state and compensate for actuator failures. Then, an online state estimation based on the FTSMC is proposed. The FTSMC is designed based on a new continuous-time sliding surface constructed to obtain an equivalent control law. Then, another arrival part of the controller is designed based on the hyperwarp algorithm. The overall structure of the FTSMC is completed by combining the super-twisting algorithm (STA)-based arrival law and equivalent control law. The final step in the controller design program is the analysis of the system dynamics and the stability of the proposed controller. A new data filter is designed for the first time to generate predictive packets using two steps, where each future control signal has a timestamp and error stamp order (ESO). The stability of the entire NCS, including the newly designed filter, is verified. Finally, the proposed method is tested on a DC motor closed-loop system. The simulation results demonstrate the advantages of the proposed control method for protecting the system from faults and disturbances and compensating for network problems.

The structure of this article is as follows. Section 6.2 introduces the robust tracking control problem of the NCS with a network disturbance equation. Section 6.3 provides the design procedure of the proposed FTSMC. In Sect. 6.4, the novel NCS method based on tracing control of the FTSMCl is described. In Sect. 6.5, the simulation results are given to verify the effectiveness of the control scheme. Section 6 provides the conclusion.

6.2 Problem Formulation

The state-space of the controlled system is expressed as follows:

$$\begin{cases} \dot{\overline{x}}(t) = \overline{A}\overline{x}(t) + \overline{B}u(t) \\ \overline{y}(t) = \overline{C}\overline{x}(t) \end{cases} \tag{6.1}$$

where $\overline{x}(t) \in R^n$ is the system state, $u(t) \in R^m$ represents the control input, and $\overline{y}(t) \in R^p$ is the output value. $\overline{A}$, $\overline{B}$, and $\overline{C}$ are known constant system matrices.

When system (6.1) is operating in a wired/wireless closed-loop environment, the system drivers and sensing components are affected by faults $\overline{f}_a(t)$ and interference $\overline{d}(t)$, as shown in Fig. 6.2, affecting actuators and sensor equipment. In Fig. 6.2, the FTSMC uses the system's estimated state $\hat{\overline{x}}$ as input. The workload of the FTSMC is the input to the NCS, and $\overline{V}_{ref}$ is a reference signal.

When system (6.1) is running through the NCS, both actuators and sensors are subject to faults and disturbances characterized by:

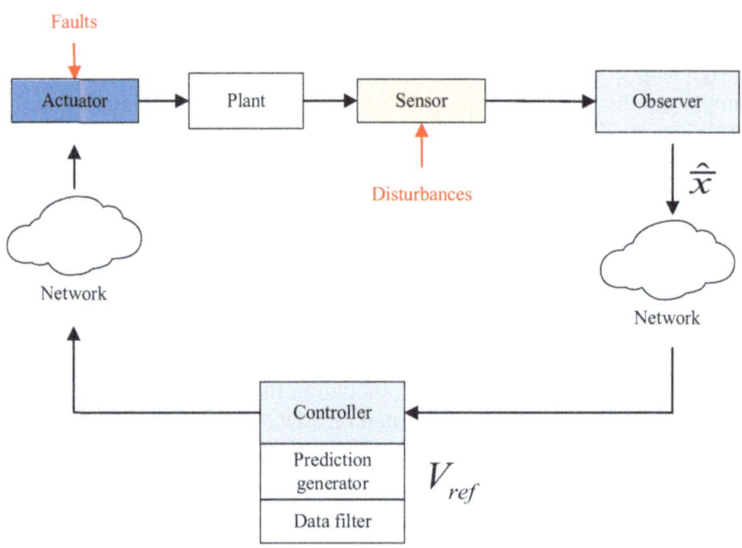

Fig. 6.2 Proposed control scheme of NCS using FTSMC under faults and disturbances

$$\dot{\overline{x}}(t) = \overline{A}\overline{x}(t) + \overline{B}u(t) + \overline{F}\overline{f}_a(t) + \overline{\omega}_a\overline{d}(t) \tag{6.2}$$

$$\overline{y}(t) = \overline{C}\overline{x}(t) + \overline{\omega}_s\overline{d}(t) \tag{6.3}$$

where $\overline{f}_a(t)$ and $\overline{d}(t)$ are the actuator and sensor faults and disturbance signals; $\overline{F}$, $\overline{\omega}_a$, and $\overline{\omega}_s$ are gain matrices.

In a closed-loop system (6.2), faults $\overline{f}_a(t)$ and disturbances $\overline{d}(t)$ are injected into the actuator, and $\overline{d}(t)$ represents disturbances of the interference system. We define the following variables to simplify the design of the protection scheme:

Let $\tilde{\overline{x}}(t) = [\overline{x}(t)\overline{\mathrm{d}}(t)]^T$, $\tilde{\overline{E}} = [I_n\, 0_{n\times p}]$, $\tilde{\overline{A}} = [\overline{A}\, 0_{n\times p}]$ and $\tilde{\overline{C}} = [\overline{C}\, I_p]$. Then, the system can be designed as:

$$\begin{cases} \tilde{\overline{E}}\dot{\tilde{\overline{x}}}(t) = \tilde{\overline{A}}\tilde{\overline{x}}(t) + \overline{B}u(t) + \overline{\omega}_a\overline{f}_a(t) \\ \overline{y}(t) = \tilde{\overline{C}}\tilde{\overline{x}}(t) \end{cases} \tag{6.4}$$

The system (6.4) will be analyzed later in this chapter.

6.3 Tracking Control Techniques

The enhanced state viewer is designed to provide the controller with dynamic system information online. Based on these online data, the FTSMC is introduced.

6.3.1 Observer Dynamic Description

We define the observer as follows:

$$\begin{cases} \dot{\overline{q}}(t) = \overline{P}\overline{q}(t) + \overline{Q}\overline{y}(t) + \overline{R}u(t) + \overline{\varpi}\gamma(t) \\ \hat{\tilde{\overline{x}}}(t) = \overline{q}(t) + \overline{S}\overline{y}(t) \end{cases} \tag{6.5}$$

where $\hat{\tilde{\overline{x}}}(t)$ is the estimated value, $\overline{q}(t)$ is the auxiliary parameter, $\overline{P}$, $\overline{Q}$, $\overline{R}$, $\overline{S}$, and $\overline{\varpi}$ are the observer parameters to be determined that $\hat{\tilde{\overline{x}}}(t)$ can converge to $\tilde{\overline{x}}(t)$; $\overline{\gamma}(t)$ is the discontinuous input to resist the actuator failure $\overline{f}_a(t)$.

We define $\overline{J} = [I_n \; -\overline{C}]^T$, and Assumption 1 is given for the observer system.

Assumption 1 The $(\overline{J}\tilde{\overline{A}}, \tilde{\overline{C}})$ pair can be detected.

In addition, we consider the following theorem to obtain the observer coefficient matrix.

Theorem 1 *Using the observer* (6.5) *of plant* (6.4), *when the fault* $\overline{f}_a(t) = 0$ *occurs, the estimation error* $\overline{\tilde{e}}(t) = \overline{\tilde{x}}(t) - \hat{\overline{\tilde{x}}}(t)$ *becomes asymptotically 0 in the absence of constraints but satisfies Assumption* 1. *In addition, the matrices* $\overline{P}$, $\overline{Q}$, $\overline{R}$, and $\overline{S}$, *which represent the observer gains, can be found.*

Proof This proof can be followed by applying a procedure similar to that in [15]; thus, it is omitted.

Remark 1 Different from the observer in [15], in [5], the equivalent injected signal method in [29] is used to construct input items $\overline{\gamma}(t)$ and actuator fault estimates $\hat{\overline{f}}_a(t)$, which will be briefly reviewed below.

We design the discontinuous switching variable $\overline{\gamma}(t)$ as follows:

$$\overline{\gamma}(t) = \overline{\rho}_{\upsilon} sign(\overline{e}_y) \tag{6.6}$$

where $\overline{e}_y = \overline{y} - \hat{\overline{y}}$, and $\overline{\rho}_{\upsilon}$ is a design scalar.

Then, the fault estimation $\hat{\overline{f}}_a(t)$ can be designed as:

$$\hat{\overline{f}}_a(t) = (\overline{\tilde{C}G}^{-1}\overline{\tilde{C}}^T\overline{F})^{\diamond}\overline{\tilde{C}G}^{-1}\overline{\tilde{C}}^T u_{eq} \tag{6.7}$$

where $\diamond$ is the general inverse, Γ and Φ are constant matrices to be designed, and υ_{eq} denotes an additional signal that can be integrated using a suitable low-pass filter as:

$$\upsilon_{eq} \cong \frac{1}{\tau s + 1}\overline{\gamma}(t) \tag{6.8}$$

where τ is a design time constant.

6.3.2 Construction of Controller

Since the FTSMC relies on the estimated $\hat{\overline{x}}(t)$ and $\hat{\overline{d}}(t)$, the state dynamics can be obtained by analyzing the second relationship of the designed observer (6.5), as follows:

$$\begin{aligned}
&\dot{\hat{\overline{x}}}(t) \\
&= \dot{\overline{q}}(t) + \overline{S}\dot{\overline{y}}(t) \\
&= \overline{P}\overline{q}(t) + \overline{Q}\overline{y}(t) + \overline{R}u(t) + \overline{\gamma}(t) + \overline{S}\dot{\overline{y}}(t) \\
&= \overline{P}(\hat{\overline{x}}(t) - \overline{S}\overline{y}(t)) + \overline{Q}\overline{y}(t) + \overline{R}u(t) + \overline{\gamma}(t) + \overline{S}\dot{\overline{y}}(t) \\
&= \overline{P}\hat{\overline{x}}(t) + (\overline{Q} - \overline{PS})\overline{y}(t) + \overline{R}u(t) + \overline{\gamma}(t) + \overline{S}\dot{\overline{y}}(t) \\
&= [\overline{P} + (\overline{Q} - \overline{PS})\tilde{\overline{C}}]\hat{\overline{x}}(t) + \overline{R}u(t) + \overline{\gamma}(k) + \overline{S\tilde{C}}\dot{\hat{\overline{x}}}(t)
\end{aligned} \tag{6.9}$$

Hence, the system state estimation can be written as:

$$\dot{\hat{\overline{x}}}(t) = (I - \overline{S\tilde{C}})^{-1}((\overline{P}(I - \overline{S\tilde{C}}) + \overline{Q\tilde{C}})\hat{\overline{x}}(t) + \overline{R}u(t) + \overline{\gamma}(t)) \tag{6.10}$$

Let $\overline{\Omega}_1 = (I-\overline{S\tilde{C}})^{-1}[\overline{P}+(\overline{Q}-\overline{PS})\tilde{\overline{C}}], \overline{\Omega}_2 = (I-\overline{S\tilde{C}})^{-1}\overline{R}$ and $\overline{\Omega}_3 = (I-\overline{S\tilde{C}})^{-1}$. It should be noted that $\hat{\overline{x}}(t) = [\hat{\overline{x}}(t)\hat{\overline{\mathrm{d}}}(t)]^T$; hence, (6.10) can be rewritten as follows to obtain $\hat{\overline{x}}(t)$:

$$\begin{aligned}
&\dot{\hat{\overline{x}}}(t) \\
&=\overline{\Omega}_1\hat{\overline{x}}(t) + \overline{\Omega}_2u(t) + \overline{\Omega}_3\overline{\gamma}(t) \\
&= \begin{pmatrix} \overline{\Omega}_{11} & \overline{\Omega}_{12} \\ \overline{\Omega}_{13} & \overline{\Omega}_{14} \end{pmatrix}\begin{pmatrix} \hat{\overline{x}}(t) \\ \hat{\overline{d}}(t) \end{pmatrix} + \begin{pmatrix} \overline{\Omega}_{21} \\ \overline{\Omega}_{22} \end{pmatrix}u(t) + \begin{pmatrix} \overline{\Omega}_{31} \\ \overline{\Omega}_{32} \end{pmatrix}\overline{\gamma}(t)
\end{aligned} \tag{6.11}$$

Based on Eq. (6.11), the following result can be obtained:

$$\dot{\hat{\overline{x}}}(t)=\overline{\Omega}_{11}\hat{\overline{x}}(t) + \overline{\Omega}_{21}u(t) + \overline{\Omega}_{31}\overline{\gamma}(t) + \overline{\Omega}_{12}\hat{\overline{d}}(t) \tag{6.12}$$

Remark 2 Based on Eq. (6.12), the estimated state $\hat{\overline{x}}(k)$ may be distorted by perturbations $\overline{d}(t)$. Therefore, the use of the term $\overline{d}(t)$ should be avoided when designing the FTSMC.

In this chapter, we define the sliding surface function as follows:

$$\overline{\sigma}(t) = \overline{e}(t) + \overline{K}_1\overline{e}(t) + \overline{K}_2\int \dot{\overline{e}}(t) \tag{6.13}$$

where $\overline{K}_1, \overline{K}_2 \in R^{1\times n}$ is designed so that $\overline{K}_1\overline{\Gamma}_{21} \neq 0.\overline{e}(t) = \hat{\overline{x}}(t) - \overline{x}_d$ is the estimation error, and $\overline{x}_d$ is the reference signal.

Remark 3 The desired system dynamics at the manifold $\overline{\sigma}(k) = 0$ is considered when the sliding surface (6.13) is designed.

Remark 4 The sliding mode function (6.13) is designed based on the NFTDSMC method. Therefore, the dynamics of the system Eq. (6.13) near the stable point of the phase is characterized by rapidly decreasing velocity.

For obtaining the sliding motion, the derivation of the sliding surface $\dot{\overline{\sigma}}(t) = 0$ is performed as follows:

$$\dot{\overline{\sigma}}(t) = (1 + \overline{k}_1 + \overline{k}_2)\dot{\overline{e}}(t) \tag{6.14}$$

where $\overline{k}_1 > 0$ and $\overline{k}_2 > 0$.

By combining Eqs. (6.12), (6.13), and (6.14), the equivalent control law can be solved as:

$$u_{eq}(t) = (\overline{\Omega}_{21})^{-1}(-\overline{\Omega}_{11}\hat{\overline{x}}(t) - \overline{\Omega}_{31}\overline{y}(t) - \overline{\Omega}_{12}\hat{\overline{d}}(t) + \dot{\overline{x}}_d(t)) \tag{6.15}$$

The reaching control law based on the STW approach u_{re} is designed as:

$$\begin{aligned} u_{re}(t) &= -r_1 \|\overline{\sigma}(t)\|^{1/2} sign(\overline{\sigma}(t)) - \overline{\Psi} + \overline{\Gamma} \\ \dot{\overline{\Psi}} &= r_2 sign(\overline{\sigma}(t)) \end{aligned} \tag{6.16}$$

where $r_1 > 2\overline{e}(t)$, $r_2 > r_1 \frac{5r_1 + 4\overline{e}(t)}{2r_1 + 4\overline{e}(t)}\overline{e}(t)$, and $\overline{\Gamma}$ is the upper bound value of the estimation error.

We combine Eqs. (6.15) and (6.16) and establish the proposed FTSMC:

$$u(t) = u_{eq}(t) + u_{re}(t). \tag{6.17}$$

6.3.3 Analysis of Proposed Algorithm

This section uses mathematical analysis to study the stability of the system using the proposed controller. We show that closed-loop system dynamics in the form of a generalized STW is possible when using the proposed controller. Then, we use the Lyapunov method and prove that the closed-loop system is asymptotically stable.

We consider the following Assumption 2 to simplify the mathematical proof.

Assumption 2 It is assumed that there is a positive number $\overline{\varphi}$ so that $\|\overline{\Gamma}\| \leq \overline{\varphi}\|\dot{\overline{\sigma}}(t)\|^{\frac{1}{2}}$.

Definition 1 Regarding a continuous-system model, the reachability of the sliding surface can be achieved once the derivative of the sliding function $\dot{\overline{\sigma}}(t) = 0$ is satisfied [27].

Theorem 2 *If the specified system* (6.12) *is combined with the controller in the slip surface* (6.13), *Eq.* (6.17), *and hypothesis 2, the system motion starting from any unstable point is forced to deflect to the slip surface* (6.12). *In a limited time, the asymptotic stability of the system dynamics can be obtained.*

Proof We substitute Eq. (6.17) into Eq. (6.14):

$$\begin{aligned}\dot{\overline{\sigma}}(t) &= (1+\overline{k}_1+\overline{k}_2)[\overline{\Omega}_{11}\hat{\overline{\dot{x}}}(t)+\overline{\Omega}_{21}((\overline{\Omega}_{21})^{-1}(-\overline{\Omega}_{11}\hat{\dot{\overline{x}}}(t)\\ &\quad -\overline{\Omega}_{31}\overline{\gamma}(t)-\overline{\Omega}_{12}\hat{\overline{d}}(t)+\dot{\overline{x}}_d(t))-r_1\|\overline{\sigma}(t)\|^{1/2}sign(\overline{\sigma}(t))\\ &\quad -\int r_2 sign(\overline{\sigma}(t))+\overline{\Gamma})+\overline{\Omega}_{31}\overline{\gamma}(t)+\overline{\Omega}_{12}\hat{\overline{d}}(t)-\dot{\overline{x}}_d(t)]\\ &= (1+\overline{k}_1+\overline{k}_2)(-r_1\|\overline{\sigma}(t)\|^{1/2}sign(\overline{\sigma}(t))-\int r_2 sign(\overline{\sigma}(t))+\overline{\Gamma})\\ &= -\overline{R}_1\|\overline{\sigma}(t)\|^{1/2}sign(\overline{\sigma}(t))-\int \overline{R}_2 sign(\overline{\sigma}(t))+\overline{\delta}\end{aligned} \tag{6.18}$$

where $\overline{R}_1 = r_1 + r_1\overline{k}_1 + r_1\overline{k}_2$, $\overline{R}_2 = r_2 + r_2\overline{k}_1 + r_2\overline{k}_2$ and $\overline{\delta} = \overline{\Gamma} + \overline{\Gamma}k_1 + \overline{\Gamma}k_2$.

We define $\overline{S}_1 = \overline{\sigma}(t)$ and $\overline{S}_2 = -\int \overline{R}_2 sign(\overline{\sigma}(t))$ so that the system dynamic (6.18) can be written as follows:

$$\begin{aligned}\dot{\overline{S}}_1 &= -\overline{R}_1\|\overline{S}_1\|^{1/2}sign(\overline{S}_1)+\overline{S}_2+\overline{\delta}\\ \dot{\overline{S}}_2 &= -\overline{R}_2 sign(\overline{S}_1)\end{aligned} \tag{6.19}$$

The system in Eq. (6.19) is the same as that in the general SWT in [30] and [31]; thus, the candidate Lyapunov functions may be similar. In this chapter, we define the Lyapunov function of system (6.19) as shown in (6.20):

$$V(\overline{S}_1) = \overline{M}^{-1}\overline{ZM} \tag{6.20}$$

where

$$\overline{Z} = \frac{1}{2}\begin{bmatrix} 4\overline{R}_2+\overline{R}_1^2 & -\overline{R}_1 \\ -\overline{R}_1 & 2 \end{bmatrix} \text{ and } \overline{M}^T = \begin{bmatrix} \|\overline{S}_1\|^2 sign(\overline{S}_1) & \overline{S}_2 \end{bmatrix}$$

Referring to definition 1, the derivative of the Lyapunov function (6.20) should be negative to confirm that the closed-loop system dynamics (6.19) will converge to zero; thus:

$$\dot{V}(\overline{S}_1) = \frac{-1}{\|\overline{S}_1\|^{1/2}}\overline{M}^T\overline{\theta M}+\frac{\overline{\varphi}}{\|\overline{S}_1\|^{1/2}}\overline{L}^T\overline{M} \tag{6.21}$$

where $\overline{\theta} = \frac{\overline{R}_1}{2}\begin{bmatrix} 2\overline{R}_2 + \overline{R}_1^2 & -\overline{R}_1 \\ -\overline{R}_1 & 1 \end{bmatrix}$ and $\overline{L}^T = \begin{bmatrix} 2\overline{R}_2 + \frac{\overline{R}_1^2}{2} & -\frac{\overline{R}_1}{2} \end{bmatrix}$ If assumption 1 is fulfilled, then the following applies:

$$\dot{V}(\overline{S}_1) \le -\frac{1}{\|\overline{S}_1\|^{1/2}}\overline{M}^T\overline{G}\overline{M} \tag{6.22}$$

where $\overline{G} = \frac{\overline{\varepsilon}_1}{2}\begin{bmatrix} 2\overline{R}_2 + \overline{R}_1^2 - 2\overline{\varphi} & * \\ -\overline{R}_1 + \frac{2\overline{\varphi}}{\overline{R}_1} & 1 \end{bmatrix}$.

We design $\begin{cases} \overline{R}_1 > 2\overline{\varphi} \\ \overline{\varepsilon}_2 > \overline{\varepsilon}_1\frac{5\overline{\varphi}\overline{R}_1 + 4\overline{\varphi}^2}{2(\overline{R}_1 - 2\overline{\varphi})} \end{cases}$ so that $\overline{G} > 0$, which implies that the derivative of the candidate Lyapunov function is negative definite. This completes the proof.

6.4 Tracking Control Law of Networked Systems

Figure 6.2 illustrates the part of the NCS system with random delays, packet data loss, and clutter. The NCS consists of two parts, the factory/actuator side and the controller side. The future signal generation device is placed on the controller side to provide the expected future control signals based on the control action to defend against communication constraints. A series of predictive control signals are then fed into the data filter to check the error value of each predictive control signal and repackage the control signals based on the error level. Assuming that the upper limit of the allowable tracking error is $\overline{e}(t) = \tilde{\Delta}$, the filter will only allow the predictive control value with the tracking error $\overline{e}(t) \prec \tilde{\Delta}$ to be transmitted through the NCS feedforward channel. As a result, the network capacity is reduced. The filter, on the other hand, arranges the output packets in ascending order according to the tracking error. A network compensator is used on the actuator side to select the appropriate control input for the factory based on the factory's time and error markers. In this paper, a round-trip time (RTT) delay τ [1, 15] and an ESO Δ_s are used to determine the control signal through the network compensator. For example, if a packet arrives at the actuator containing multiple signals with the same ESO, the compensator will use a timestamp to select the appropriate signal. In other words, if the data packet has a network time delay due to packet dropout and packing disorder, the compensator will admit ascending ESO stamp. The following two assumptions are made.

Assumption 3 The controllers and actuators depend on events, whereas the sensors depend on time. Each node is clock-synchronized.

Assumption 4 The RTT τ is bounded, i.e., $0 \le \tau \le \pi$, and the ESO Δ_s is also bounded at $\tilde{\Delta}$ with an ascending sequence so that $\Delta_1 \le \Delta_2 \le \ldots\ldots \le \Delta_i \le \tilde{\Delta}$.

Remark 5 On the actuator side, the RTT represents the total delay occurring in the NCS, which can be obtained by simply comparing the timestamp to the latest control batch of the existing timestamp on the actuator at each sampling time. This dealy represents all delay times occurring through the network feedback and the feedforward channels due to network disturbances.

6.4.1 Actuator/Plant Side

Sensor: An observer (6.5) is designed to obtain factory packet information based on sensor measurements; it sends these packets with a timestamp t_s through the network feedback channel:

$$\overline{H}_{t_s} = \left\{\hat{\overline{x}}(t), \hat{\overline{d}}(t), \overline{\gamma}(t)\right\} \tag{6.23}$$

Network compensator: The network compensator receives the packet information through the network feed forward channel with RTT and ESO tags. The compensator performs the following updates at each sample time T_s:

$$u_{t_s,\Delta_s} = \begin{cases} u(\Delta_i, t_i), \ t_i \leq t_s \\ u(\Delta_i, t_i), \ t_i \geq t_s \\ u(t_i, \Delta_i), \ \Delta_{i-1} = \Delta_i = \Delta_{i+1} \end{cases} \tag{6.24}$$

Remark 6 In Eq. (6.24), $u(\Delta_i, t_i)$ denotes that the decision priority of the compensator selection is based on the error stamp Δ_s, and $u(t_i, \Delta_i)$ denotes that the time stamp t_s RTT has been included.

6.4.2 Controller Side

Prediction Generator: The main purpose of this device is to receive feedback packets. It then predicts the system state and controls the inputs according to Eqs. (6.12), (6.15), and (6.16) until the time $t_s + \tau$ is as follows:

$$\dot{\hat{\overline{x}}}(t)=\overline{\Omega}_{11}\hat{\overline{x}}(t) + \overline{\Omega}_{21}u(t) + \overline{\Omega}_{31}\overline{\gamma}(t) + \overline{\Omega}_{12}\hat{\overline{d}}(t) \tag{6.25}$$

$$\begin{aligned} u(t) = {} & (\overline{\Omega}_{21})^{-1}(-\overline{\Omega}_{11}\hat{\overline{x}}(t) - \overline{\Omega}_{31}\overline{\gamma}(t) - \overline{\Omega}_{12}\hat{\overline{d}}(t) + \dot{\overline{x}}_d(t)) \\ & - r_1\|\overline{\sigma}(t)\|^{1/2} sign(\overline{\sigma}(t)) - \int r_2 sign(\overline{\sigma}(t)) + \overline{\Gamma} \end{aligned} \tag{6.26}$$

According to Eqs. (6.25) and (6.26), the predictive generator produces the following packet:

$$\overline{L}_{t_s} = \left[u(t_1|t_s)^T, u(t_2|t_s)^T, \ldots, u(t_s + \tau|t_s)^T\ \right]^T \tag{6.27}$$

Then $\overline{L}_{t_s}$ is grouped simultaneously with the time stamp t_s and passed on to the data filter.

Data filters: Data filters are designed to achieve two main functions, namely, network capacity savings and strong tracking control performance. From Eq. (6.27) onwards, regardless of the data filter, all predictive control signal series will be sent through the feedforward channel independent of the final error level of the control signal. The following Luenberger observer is designed to obtain future errors due to specific control signals.

$$\dot{\hat{\overline{x}}}(t_s) = \overline{\varphi}\hat{\overline{x}}(t_s) + \overline{\Theta}u(t_s) + \overline{w}(\overline{y}(t_s) - \tilde{\overline{C}}\hat{\overline{x}}) \tag{6.28}$$

$$\hat{\overline{y}}(t_s) = \tilde{\overline{C}}\hat{\overline{x}}(t_s) \tag{6.29}$$

where $\overline{w}$ is the Luenberger gain that will be determined, and $\hat{\overline{y}}(t_s)$ is the observer output.

We define the future tracking error as:

$$\overline{\wp} = \hat{\overline{y}}(t_s) - \overline{y}_d(t_s) \tag{6.30}$$

where $\overline{y}_d(t_s)$ is the desired reference signal.

The data filter output packet is sent through the feed-forward channel with RTT and ESO stamps as follows:

$$\overline{Y}_{(t_s,\Delta_s)} = \left[u(t_1, \Delta_1|t_s, \Delta_s)^T, u(t_2, \Delta_2|t_s, \Delta_s)^T, \ldots, u(t_s + \tau, \Delta_s|t_s, \Delta_s)^T\ \right]^T \tag{6.31}$$

where t_s and Δ_s are the time and error stamp, respectively.

Remark 7 Unlike in [1] and [15], according to Eq. (6.31), this study proposes a double stamp used with the data packet to facilitate the compensation selection decision and save network resources.

6.4.3 Analysis of System Dynamics Stability

According to Assumption (6.2) and the network compensation update selection (6.24), the control signal at the actuator input at a specific sample time instance t_s can be described as follow:

$$u_{t_s} = u(t_s + \tau, \Delta_s | t_s, \Delta_s) \tag{6.32}$$

We substitute Eq. (6.32) in the first equation of (6.2), and replace $\hat{\overline{x}}$ with the estimated value instead of $\overline{x}$ in Eq. (6.2). The following closed-loop NCS can be obtained:

$$\dot{\hat{\overline{x}}}(t_s) = \overline{A}\hat{\overline{x}}(t_s) + \overline{B}(u(t_s + \tau, \Delta_s | t_s, \Delta_s) + \overline{\Upsilon}) \tag{6.33}$$

Since $\overline{\Upsilon} = \hat{\overline{f}}_a(t) + \hat{\overline{d}}(t)$, we denote $V(t) = [\hat{\overline{x}}_1(t) \,..\, \hat{\overline{x}}_\ell(t)]^T \in R^\ell$ as a state vector that expresses the closed-loop system dynamics at a random time t; then:

$$\dot{V}(t) = \overline{M}(t)V(t) + \overline{\beta}(t)(u(t) + \overline{\Upsilon}) \tag{6.34}$$

If $\overline{M} \in R^y$, there exists a nonsingular matrix $\overline{\Im}$ that satisfies Eq. (6.34):

$$\overline{\Im\beta} = \begin{bmatrix} 0_{(\ell-y)xy} \\ \overline{\beta}_s \end{bmatrix} \tag{6.35}$$

where $\overline{\beta}_s \in R^{yxy}$ is nonsingular. We denote $\overline{\zeta} = \overline{\Im}V(t)$, and the dynamic Eq. (6.33) can be rewritten as follows:

$$\dot{\overline{\zeta}}(t) = \tilde{\overline{\Omega}}\overline{\zeta}(t) + \begin{bmatrix} 0_{(\ell-y)xy} \\ \overline{\beta}_s \end{bmatrix}(u(t) + \overline{\Upsilon}) \tag{6.36}$$

where $\tilde{\overline{\Omega}} = \overline{\Im M \Im}^{-1}$, and the system (6.36) can be rewritten as:

$$\dot{\overline{\zeta}}_1(t) = \tilde{\overline{\Omega}}_{11}(t)\overline{\zeta}_1(t) + \tilde{\overline{\Omega}}_{12}(t)\overline{\zeta}_2(t) \tag{6.37}$$

$$\dot{\overline{\zeta}}_2(t) = \tilde{\overline{\Omega}}_{21}(t)\overline{\zeta}_1(t) + \tilde{\overline{\Omega}}_{22}(t)\overline{\zeta}_2(t) + \overline{\beta}_2(t)(u(t) + \overline{\Upsilon}) \tag{6.38}$$

It can be seen that system dynamic (6.37) expresses the dynamics of the NCS (6.36) [1] and [15]. According to Eqs. (6.14), (6.30), and (6.34) the relations between the state vector $V(t)$, tracking error $\overline{\wp}$, and the sliding surface $\overline{\sigma}(t)$ can be written as Eqs. (6.39) and (6.40):

$$\overline{\sigma}(t) = f(.)\overline{\wp} \tag{6.39}$$

$$\overline{\wp} = g(.)V(t) \tag{6.40}$$

where $f(.)$ and $g(.)$ are the designed matrices.

By substituting Eq. (6.39) into Eq. (6.40), we obtain:

$$\overline{\sigma}(t) = \overline{G}V(t) \tag{6.41}$$

where $\overline{G} = f(.).g(.)$. Hence, the sliding mode function is defined in Eq. (6.42):

$$\begin{aligned}\overline{\sigma}(t) &= \overline{G}V(t) = \overline{G}(\overline{\Im})^{-1}\overline{\zeta} \\ &= \overline{\tilde{G}}\overline{\zeta} = [\overline{\tilde{G}}_1\ \overline{\tilde{G}}_2][\overline{\zeta}_1(t)\ \overline{\zeta}_2(t)\ ]\end{aligned} \tag{6.42}$$

where $\overline{\tilde{G}} \in R^{\ell x(y-\ell)}$ is nonsingular.

When the sliding manifold is reached i.e., $\overline{\sigma}(t) = 0$, then

$$\overline{\tilde{G}}_1\overline{\zeta}_1(t) + \overline{\tilde{G}}_2\overline{\zeta}_2(t) = 0 \tag{6.43}$$

Substituting $\overline{\zeta}_2(t) = -\overline{\tilde{G}}_2^{-1}\overline{\tilde{G}}_1\overline{\zeta}_1(t)$ into Eq. (6.37) yields:

$$\begin{aligned}\dot{\overline{\zeta}}_1(t) &= \overline{\tilde{\Omega}}_{11}(t)\overline{\zeta}_1(t) + \overline{\tilde{\Omega}}_{12}(t)\overline{\tilde{G}}_2^{-1}\overline{\tilde{G}}_1\overline{\zeta}_1(t) \\ &= (\overline{\tilde{\Omega}}_{11}(t) + \overline{\tilde{\Omega}}_{12}(t)\overline{\tilde{G}}_2^{-1}\overline{\tilde{G}}_1)\overline{\zeta}_1(t)\end{aligned} \tag{6.44}$$

Theorem 3 *The asymptotic stability of the deduced system* (6.44) *depends on the existence of a positive-definite matrix with symmetric dimension* $\overline{\tilde{\varpi}} \in R^{(y-\ell)x(y-\ell)}$ *so that*:

$$\begin{bmatrix} -\overline{\tilde{\varpi}} & (\overline{\tilde{\Omega}}_{11}(t) + \overline{\tilde{\Omega}}_{12}(t)\overline{\tilde{G}}_2^{-1}\overline{\tilde{G}}_1)^T \\ (\overline{\tilde{\Omega}}_{11}(t) + \overline{\tilde{\Omega}}_{12}(t)\overline{\tilde{G}}_2^{-1}\overline{\tilde{G}}_1) & \overline{\tilde{\varpi}}^{-1} \end{bmatrix} \prec 0 \tag{6.45}$$

Proof We construct the following Lyapunov function:

$$V(t) = \overline{\zeta}_1(t)\overline{\tilde{\varpi}}\overline{\zeta}_1(t) \tag{6.46}$$

According to the Lyapunov function, the derivative of $V(t)$ along the trajectory of the system (6.44) yields:

$$\begin{aligned}\dot{V}(t) &= \dot{\overline{\zeta}}_1^T(t)\tilde{\overline{\varpi}}\overline{\zeta}_1(t) + \overline{\zeta}_1^T(t)\tilde{\overline{\varpi}}\dot{\overline{\zeta}}_1(t) \\ &= \overline{\zeta}_1^T(t)(\tilde{\overline{\Omega}}_{11}(t) + \tilde{\overline{\Omega}}_{12}(t)\tilde{\overline{G}}_2^{-1}\tilde{\overline{G}}_1)^T\tilde{\overline{\varpi}}(\tilde{\overline{\Omega}}_{11}(t) \\ &\quad + \tilde{\overline{\Omega}}_{12}(t)\tilde{\overline{G}}_2^{-1}\tilde{\overline{G}}_1)\overline{\zeta}_1(t) - \overline{\zeta}_1^T(t)\tilde{\overline{\varpi}}\overline{\zeta}_1^T \prec 0\end{aligned} \tag{6.47}$$

Hence,

$$(\tilde{\overline{\Omega}}_{11}(t) + \tilde{\overline{\Omega}}_{12}(t)\tilde{\overline{G}}_2^{-1}\tilde{\overline{G}}_1)^T\tilde{\overline{\varpi}}(\tilde{\overline{\Omega}}_{11}(t) + \tilde{\overline{\Omega}}_{12}(t)\tilde{\overline{G}}_2^{-1}\tilde{\overline{G}}_1) - \tilde{\overline{\varpi}} \prec 0 \tag{6.48}$$

The matrix inequality (7–48) can be obtained by applying the Schur complement. This finishes the proof.

6.5 Simulation Results

We consider the DC motor model with a state-space representation as defined in Eq. (6.2) where.
$\overline{A} = \begin{bmatrix} -2.219\ 4.438 \\ -12.000\ \text{-}\ 24.000 \end{bmatrix}$, $\overline{B} = \begin{bmatrix} 0 \\ 20 \end{bmatrix}$, $\overline{C} = \begin{bmatrix} 1\ 0 \\ 0\ 1 \end{bmatrix}$, $\overline{F} = \begin{bmatrix} 0.001 \\ 19.999 \end{bmatrix}$, $\overline{\omega}_a = \begin{bmatrix} 0.1000 \\ 0.1000 \end{bmatrix}$ and $\overline{\omega}_s = \begin{bmatrix} 1 \\ 0 \end{bmatrix}$.

The parameters of the observer are:

$$\overline{P} = \begin{bmatrix} -3.6633 & 3.5496 & -0.8963 \\ -6.4111 & -13.8227 & 0.0534 \\ 1.7734 & -5.3285 & -0.0963 \end{bmatrix}, \overline{R} = \begin{bmatrix} 0.7406 \\ 10.6850 \\ 0.7425 \end{bmatrix}, \overline{Q} = \begin{bmatrix} 0.1037 & 3.1067 \\ 0.0534 & -5.2902 \\ -0.0963 & -4.0504 \end{bmatrix},$$

$$\overline{S} = \begin{bmatrix} 0.8960 & -26.5286 \\ -1.4081 & 53.9909 \\ -0.0914 & 34.0882 \end{bmatrix}, \overline{\varpi} = \begin{bmatrix} 0.0378 & 0 \\ 0 & 0.0378 \\ 0.0378 & 0 \end{bmatrix}.$$

The controller parameters are $\overline{K}_1 = 1$, $\overline{k}_2 = 0.8$, $r_1 = 5$, and $r_2 = 6$. The Luenberger observer gain and RTT values are the same as in [1] and [6].

In the simulations, the square wave signal is chosen as the reference signal, with a range between −4 and 4 r/s (to represent the forward and reverse directions of the motor) at time t = 0–8 s. At 8–15 s, the speed ranges from −7 to 7 r/s. The simulation time is 15 s, and the DC motor system is affected by the following disturbances and faults:

$$\overline{\Psi}(t) = \begin{cases} 0.08\sin(2\pi t), & 0 \le t \le 10s \\ 6\sin(5\pi t) + [0.10.5]x, & 10s < t < 15s \end{cases}$$

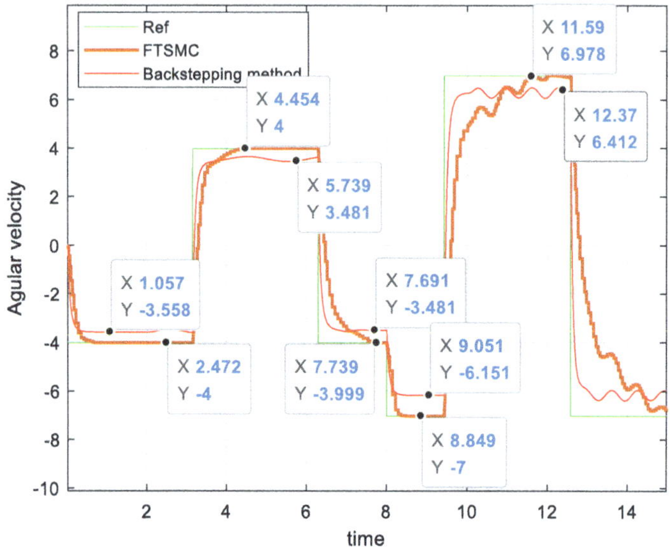

Fig. 6.3 Angular velocity differentiable fault case

$$\overline{\Phi}(t) = \begin{cases} 0, & 0s \le t \le 2s \\ 0.08(t-2)^2 + 2\sin(\pi(t-2)), & 2s < t < 7s \\ 1.8, & 7s < t < 10s \\ 3\sin(6\pi(t-10)) + 1.2, & 10s < t \le 15s \end{cases}$$

The proposed FTSMC method is compared with the backstepping control in [29]. The results are shown as follows (Fig. 6.3).

The reference speed is $\overline{x}_d = 4$ r/s with a range of 0–8 s. The proposed controller angular velocity is 4 r/s, the same as the reference speed. For the backstepping method in [29], the nearest value is 3.56 r/s in the other direction, while the FTSMC controller tracking speed is 4 r/s at t = 4.454 s, and that of the other controller is 3.48 r/s at t = 5.739 s. At t = 8 s, the reference speed changes to 7 r/s. The motor angular speed for the proposed controller is 7 rad/sec at t = 8.849 s. The backstepping controller is 6.151 r/s at the same time. In the other direction, the proposed method can track the reference speed while the compared method is 6.412 r/s at t = 12.37 s, indicating a large tracking error. We can conclude from this analysis that the proposed controller is robust, especially during transients, even though the system experiences faults, disturbances, and network constraints.

Figure 6.4 shows the control performance of the proposed method and the method in [29]. The overshoot ratio of the backstepping controller signal in the transient process is 181 and 182.9, whereas the control signal value of the proposed controller is 33.44 at t = 3.19 s and 33.74 at t = 6.283 s, indicating that the proposed method provides power savings. In addition, the overshoot ratio of the proposed controller is lower than that of the backstepping control method in [29].

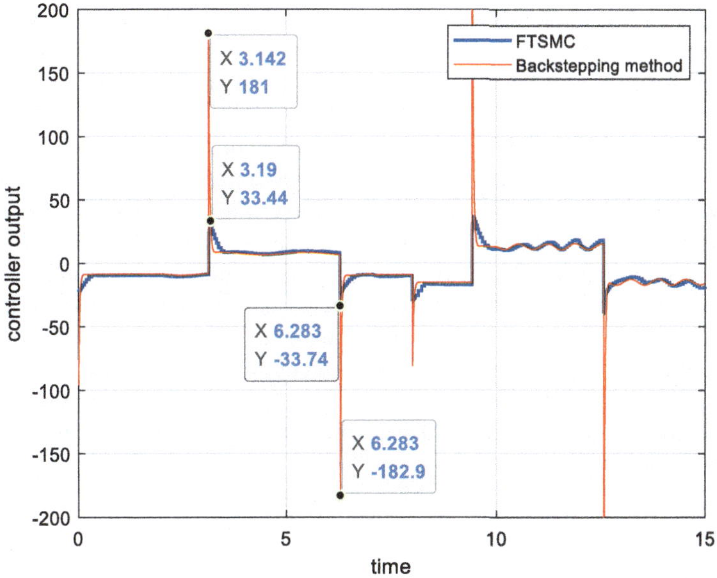

Fig. 6.4 Control output differentiable fault case

Figure 6.5 shows the tracking error of both control methods. During the transient periods of the two controllers, the proposed FTSMC method has a smaller tracking error of 0.0001295 and 0.000304 r/s at t = 2.626 s and t = 8.678 s. In contrast, the

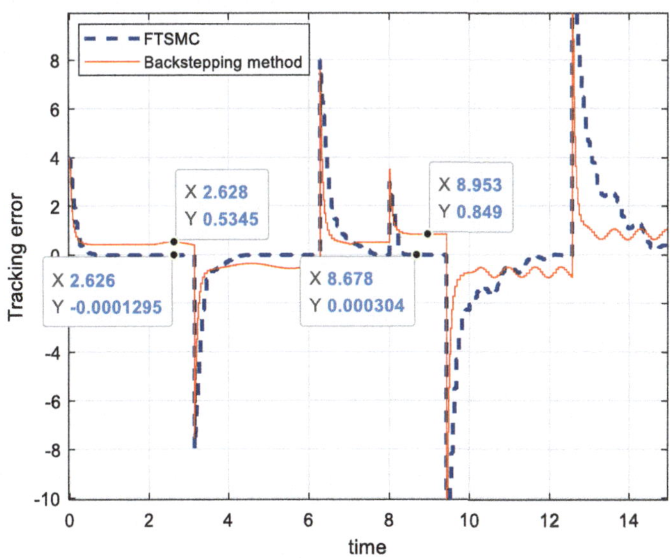

Fig. 6.5 Tracking error differentiable fault case

backstepping control method has a larger tracking error of 0.5345 and 0.849 at t = 2.628 s and t = 8.953 s. The simulation results demonstrate that the FTSMC effectively compensates for system faults and reduces system disturbances and network constraints.

In this chapter, faults, interferences, and network constraints in a network control model were investigated. The FTSMC was proposed to compensate for failures and minimize actuator and sensor disturbances. First, a long-term state observer was constructed to provide the state and defend against system failures online. The FTSMC was established based on instantaneous estimation information. A new continuous-time sliding function was constructed using the nonsingular fast terminal method. In addition, the proposed control method was analyzed mathematically, demonstrating the superiority of the proposed control scheme. Finally, a DC motor system was used to test the proposed technology. The simulation results showed that the proposed FTSM controller could deal with faults and disturbances in the network system efficiently.

References

1. M. Li and Y. Chen, "Robust tracking control of networked control systems with communication constraints and external disturbance," IEEE Transactions on Industrial electronics, vol. 64, no. 5, pp. 360-372, May 2017.
2. M. Li and Y. Chen, "Challenging research for networked control systems-a survey," Transactions of the Institute of Measurement and Control, vol. 41, no. 9, pp. 2400-2418, 2019.
3. K. Liu, E. Fridmanb, and K. Johansson, "Dynamic quantization of uncertain linear networked control systems," Automatica, vol.59, pp.248-255, Sep, 2015.
4. Y. Xia, "Cloud control systems," IEEE/CAA J. Autom. Sinica,, vol.2,no 2, pp.134-142, Apr. 2015.
5. M. Cho and C. Lee, "A low-power real-time operating system for ARC (actual remote control) wearable device," IEEE, Consum..Electron. vol.56, no. 3, pp.1602-1609, Aug.2010.
6. M. Li and Y Chen, "Sliding Mode Predictive Tracking Control for DC Permanent Magnet Motor in NCSs with Random Delay and Packet Dropouts," International Journal of Control, Automation and Systems, 15(6): 2723-2732., 2017.
7. Li M and Chen Y, "Wide-area stabiliser on sliding mode control for cross-area power systems with random delay and packet dropouts," Asian Journal of Control, vol. 20, no. 6, pp. 1–13, Nov. 2018.
8. W. Chengwei, L. Jianxing, X. Yongyang, and W. Ligang, "Observer-based adaptive fault-tolerant tracking control of nonlinear Nonstrict-Feedback Systems," IEEE Transactions on Neural Networks and Learning Systems. vol. 29, no. 7, pp.3022-3033, 2018.
9. J. Wang and G. Yang, "Data-driven output-feedback fault-tolerant tracking control method and its application to a dc servo system," in IEEE/ASME Transactions on Mechatronics, vol. 24, no. 3, pp. 1186-1196, Jun. 2019.
10. G. Franze, F. Tedesco, and D. Famularo, "Model predictive control ' for constrained networked systems subject to data losses," Automatica, vol. 54, pp. 272–278, Apr. 2015.
11. R. Wang, B. Wang, G. Liu, W. Wang, and D. Rees, "H∞ controller design for networked predictive control systems based on the average dwell-time approach," IEEE Transactions Circuits Syst. II, Exp. Briefs, vol. 57, no. 4, pp.310–314, Apr. 2010.

12. Y. Liu and I. Fong, "Robust predictive tracking control of networked control systems with time-varying delays and data dropouts." IET Control Theory Application, vol. 7, no. 5, pp. 738–748, May 2013.
13. A. Loza, J. Cieslak, D. Henry, A. Zolghadri, and L. Fridman, "Output tracking of systems subjected to perturbations and a class of actuator faults based on hosm observation and identification," Automatica, vol. 59, pp.200–205, 2015.
14. A. Teixeira, J. Araújo, H. Sandberg, and K. Johansson, "Distributed sensor and actuator reconfiguration for fault-tolerant networked control systems" IEEE Transactions on Control of Network Systems, vol. 5, no. 4, pp.1517-1528, Dec. 2018.
15. M. Li and Y. Chen, "Robust adaptive sliding mode control for switched networked control systems with disturbance and faults," IEEE Transactions on Industrial Informatics, vol. 15, no. 1, pp. 193 - 204, Jan. 2019.
16. G. P. Liu, Y. Q. Xia, J. Chen, D. Rees, and W. S. Hu, "Networked predictive control of systems with random network delays in both forward and feedback channels," IEEE Transactions on Industrial Electronics, vol. 54, no. 3, pp. 1282-1297, Jun. 2007.
17. R. Q. Lu, Y. Xu, and R. D. Zhang, "A new design of model predictivetracking control for networked control system under random packet loss and uncertainties," IEEE Transactions on Industrial Electronics, vol. 63, no. 11, pp. 6999-7007, Nov. 2016.
18. X. M. Sun, Di Wu, G. P. Liu, and W. Wang, "Input-to-state stability for networked predictive control with random delays in both feedback and forward channels," IEEE Transactions on Industrial Electronics, vol. 61, no. 7, pp. 3519–3526, Jul. 2014.
19. Y. Chen and B. Guo, "Sliding mode fault tolerant tracking control for a single-link flexible joint manipulator system" IEEE Access, vol. 31, Jul. 10, 2019.
20. Amin, A. Ahmed, Hasan, and K. Mahmood, "A review of fault tolerant control systems: advancements and applications," Measurement, vol.143, pp.58–68, 2019.
21. V. Utkin, "Variable structure systems with sliding modes," IEEE Transactions on Automatic Control, vol. 22, no. 2, pp. 212-222, Apr. 1977.
22. R. Guzman, L.G. Vicu˜na, J. Morales, M. Castilla, and J. Matas, "Sliding-mode control for a three-phase unity power factor rectifier operating at fixed switching frequency," IEEE Transactions on Power Electronics, vol. 31, no. 01, pp. 758–769, Jan. 2016
23. S. Qu, X. Xia, and J. Zhang, "Dynamics of discrete-time sliding-mode-control uncertain systems with a disturbance compensator," IEEE Transactions on Industrial Electronics, vol. 61, no. 7, pp. 3502- 3510, Jul. 2014
24. S. Li, C. Wu, and Z. Sun, "Design and implementation of clutch control for automotive transmissions using terminal-sliding-mode control and uncertainty observer," IEEE Transactions on Vehicular Technology, vol. 65, no. 4, pp. 1890-1898, Apr. 2016
25. W. Gao, Y. Wang, and A. Homaifa, "Discrete-time variable structure control systems," IEEE Transactions on Industrial Electronics, vol. 42, no. 2, pp. 117-122, Apr. 1995
26. J. Davila, L. Fridman, and A. Levant, "Second-order sliding-mode observer for mechanical systems," IEEE Transactions on Industrial Informatics, vol. 15, no. 1, pp. 193 - 204, Jan. 2019
27. M. Van, S. Ge, and H. Ren, "Finite time fault tolerant control for robot manipulators using time delay estimation and continuous nonsingular Fast Terminal Sliding Mode Control," IEEE Transactions on Cybernetics, vol. 47, no. 7, pp. 1681 - 1693, Jul. 2017
28. B. Veselí, B. P. Draženovic, and C. Milosavljevic,, "Improved discrete-time sliding-mode position control usingEuler velocity estimation," IEEE Transactions on Industrial Electronics, vol. 57, no. 11, pp. 3840-3847, Nov. 2010
29. J. Lan and R. Patton, "A decoupling approach to integrated fault-tolerant control for linear systems with unmatched non-differentiable faults," Automatica, Vol.89, pp.290-299, Mar. 2018
30. M. Van, H. Kang, and K. Shin, "Novel quasi-continuous super-twisting high-order sliding mode controllers for output feedback tracking control of robot manipulators," Journal of Mechanical Engineering Science., vol. 228, no. 17, pp.3240-3257, 2014
31. Khalil and M. Osorio, "A Lyapunov approach to second-order sliding mode controllers and observers," in Proc. 47th IEEE Conf. Decis. Control, Cancún. Mexico, pp. 2856–2861, 2008

Chapter 7
Super-Twisting Sliding-Mode Tracking Control for Networked Systems with Multi-Channels Transmission

This chapter proposes a second-order super-twisting sliding mode control (SSTSMC) method to address the shortcomings of speed tracking control in multi-transmission channels and remote control systems of a DC motor system. First, the multi-channel and external disturbance DC motor model is considered. Then, an observer is proposed in the form of a super-twisting algorithm (STA) to estimate the state of the system and suppress jitter. In addition, the SSTSMC algorithm based on the second-order super-twisting observer (STO) is designed to track the speed of the DC motor system. A nonlinear term is used in the controller to suppress external disturbances and jitter when switching between channels. The stability of the control method is demonstrated. Finally, simulations are carried out to prove the applicability of the control method.

7.1 Introduction

DC motors have good speed regulation performance and dynamic characteristics and have been widely used in satellites, aerospace, robot control, and other fields [1]. With the rapid development of military, medical, and industrial control applications, the requirements of DC motor speed regulation accuracy and response speed have increased in recent years, attracting widespread attention from scholars. In addition, in modern industrial control systems, the parameters of the controlled object are typically obtained from multiple sensors and transmitted through different channels [2].

Switching between sensors and channels inevitably causes abrupt changes in the parameters; thus, the Markov jump theory can be effectively used in these systems. In [3], a hidden Markov model was proposed to deal with the possible asymmetry between the system mode and the controller mode and ensure that the closed-loop system followed a hidden Markov motion. In [4], a robust controller for a network control system for multiple communication channels was developed, and a Markov

© The Author(s), under exclusive license to Springer Nature Singapore Pte Ltd. 2022
M. Li et al., *Tracking Control of Networked Systems via Sliding-Mode*,
https://doi.org/10.1007/978-981-16-6514-1_7

channel switching algorithm was used. In [5], an equivalent control method for multi-input multi-output (MIMO) uncertain linear Markov jump systems based on sliding mode control (SMC) was designed to ensure that the system was stochastically asymptotically stable. Therefore, it is theoretically possible to employ Markov hopping to for switching between multiple channels.

Various advanced control algorithms have been proposed; examples include, proportional–integral–derivative (PID) control [6], H-infinity control [7, 8], fuzzy control [9, 10], model predictive control [11, 12], event-triggered control [13, 14], and SMC [15, 16]. SMC has the advantages of a simple algorithm, strong robustness, and fast response speed. It prevents uncertainty of the system and is suitable for nonlinear systems. Some scholars applied the SMC method to space robot control. SMC is an important nonlinear control method and has been extensively studied in recent years. However, SMC has several shortcomings. When the trajectory of SMC is close to the sliding surface, it usually moves back and forth within the sliding mode bandwidth, leading to a movement to the equilibrium point, which is a substantial problem in practical applications.

Many studies focused on methods for suppressing jitter, including high-order SMC and adaptive SMC. Xu [16] proposed an SMC strategy based on a terminal sliding surface and designed a new secondary sliding surface. A second-order discrete-time SMC strategy was used for the precise motion control of piezoelectric-driven nano-positioning devices. By using the feedback of the output value, this strategy was easy to implement and did not require the use of a state observer. In [17], an integral suboptimal second-order SMC algorithm was proposed. The controller adjusted by this algorithm differed from the traditional SMC, which provides continuous control to the system; thus, it is more suitable for the operation of industrial robots. The algorithm has a positive damping effect and was demonstrated for use in the industry. Merabet et al. [18] designed an SMC method to control the wind energy conversion between the generator and the grid. This method is based on an external interference model and has strong robustness. The experiments showed that the method was robust to uncertain interference, parameter changes, and system uncertainty.

The above method is effective for controlling the system. However, chattering is not sufficiently suppressed, and the method cannot be quickly adapted to other areas. Therefore, some scholars suggested using nonlinear terms in control strategies; the STA is a popular method. This term effectively and quickly eliminates jitter near the sliding surface so that the system can be stabilized faster than when a conventional SMC is used. For example, in [19], a continuous third-order integral terminal sliding mode control (3-ITSMC) method based on STA was proposed to track the motion of a piezoelectric-driven nano-positioning system. This strategy eliminates jitter and produces a higher sliding mode accuracy. The proposed 3-ITSMC performed better than the traditional second- and third-order SMC control methods in terms of robust tracking, external interference suppression, and other problems caused by uncertain parameters. In [20], Moreno et al. give the theory of constructing a strict Lyapunov function, which ensured that the proposed sliding surface was accessible and the system could stabilize in a limited number of steps. The proposed STA-based control strategy significantly reduced the jitter problem caused by the conventional SMC. However, the STA-based controller method has not been widely analyzed and applied. For example, few papers described the speed tracking control of STA-based DC motor systems with uncertain parameters and external disturbances. Although a high-end SMC can effectively reduce jitter, its structure is complex and difficult to implement or cannot be used in the industry. Therefore, it is crucial to investigate this method, especially the SSTSMC strategy.

After reviewing the above literature, a DC motor system model with switching channels and external interference is established. The channel switching follows a Markov jump process. An STO is proposed to eliminate system status errors caused by switching channels and external interference. Based on the STO, the SSTSMC algorithm is designed to track the speed of the DC motor. A nonlinear term in the form of an STA is used in the algorithm to reduce the jitter caused by switching channels. Finally, the proposed control method is analyzed using numerical simulations and experiments. The results illustrate the effectiveness of the proposed method.

This chapter is organized as follows. In Sect. 7.2, the speed tracking control problem of the DC motor system is explained. We consider a system with multiple channels and external interference where the channel switching follows a Markov jump process. Section 7.3 describes the second-order super-twisting controller based on the STO. Section 7.4 presents the numerical simulations to demonstrate the usability of the proposed method. Section 7.5 concludes the chapter.

7.2 Problem Formulation

We design the state-space model of the DC motor system and consider the switching of multiple transmission channels and external interference that may occur in network control scenarios. The structure of the DC motor with switching channels and external interference is shown in Fig. 7.1. The system consists of a controller and a DC motor device. The controller calculates the appropriate control value based on the measurement signal (digital value). After converting the data from digital to analog, the value is transmitted to the DC motor through the transmission channel, and multiple transmission channels are used. Each channel has different channel characteristics and transmission errors. It is assumed that the channel selection is a random process, and jitter is a common occurrence.

The working state-space equation of the DC motor system can be expressed as [22]:

$$\tilde{x}(k+1) = \overline{A}\tilde{x}(k) + \overline{B}u(k) + \gamma(k) \tag{7.1}$$

$$\tilde{\zeta}(k) = \tilde{y}_i(k) = \overline{C}_i\tilde{x}(k) \tag{7.2}$$

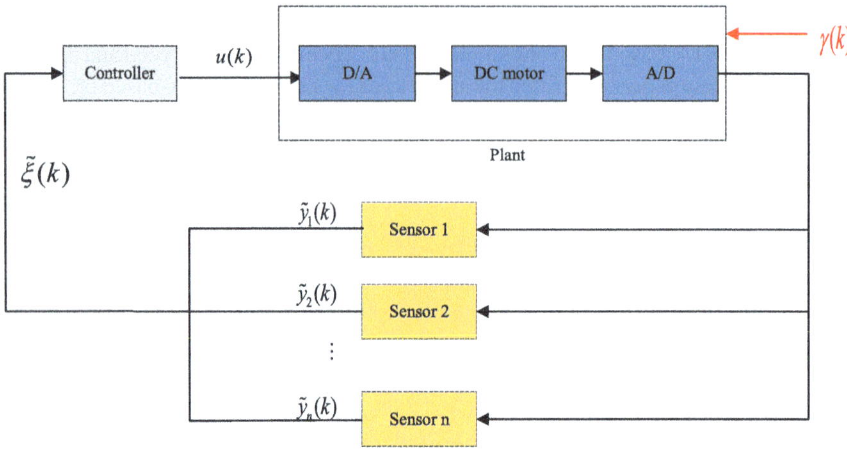

Fig. 7.1 DC motor system with multi-channels and external disturbances

where $\tilde{x}(k)=[\tilde{x}_1(k)\,\tilde{x}_2(k)]^T$, $\tilde{x}_1(k)$, $\tilde{x}_2(k) \in R$, $u(k) \in R$, and $\tilde{y}_i(k) \in R$ are the system states, control input, and output value, respectively. Meanwhile, $\gamma(k) = \left[\gamma_1(k)\,\gamma_2(k)\right]^T$, $\gamma_1(k)$, and $\gamma_2(k) \in R$ represent the external disturbances. A, B is the system's parameter-dependent matrix with appropriate dimensions. $\overline{C}_i$ is the constant matrix with an appropriate number of channels i. The differences between $\overline{C}_1, \overline{C}_2, \ldots, \overline{C}_n$ are set to simulate the different errors and measurement noises between channels.

According to the stochastic communication protocol (SCP) [21], only one channel $\tilde{\xi}(k) \in \{\tilde{y}_1(k), \tilde{y}_2(k), \ldots, \tilde{y}_n(k)\}$ is chosen to transmit data at one time.

Assumption 1 The selection of the channel $\tilde{\xi}(k)$ follows a Markov jump process.

Based on Assumption 1, the probability of $\tilde{\xi}(k+1) = j$ when $\tilde{\xi}(k) = i$ can be formulated as follows:

$$\text{Prob}\{\tilde{\xi}(k+1) = j|\tilde{\xi}(k) = i\} = \pi_{ij}(k) \tag{7.3}$$

where $\pi_{ij}(k) \geq 0, (i, j \in \{1, 2, \ldots, n\})$ denotes the conditional transition probability from $\tilde{\xi}(k) = i$ to $\tilde{\xi}(k+1) = j$, which satisfies:

$$\sum\nolimits_{j=1}^{n} \pi_{ij}(k) = 1, (i \in \{1, 2, \ldots, n\}) \tag{7.4}$$

7.3 Establishment of the Super-Twisting Tracking Strategy

7.3.1 Design of Observer

A sliding mode observer based on the second-order STA is proposed to suppress external disturbances and observe the state variables. In the state-space equation, we define the error of the system state variable and the measurements as follows:

$$\begin{cases} \tilde{e}_1(k) = \hat{\tilde{x}}_1(k) - \tilde{x}_1(k) \\ \tilde{e}_2(k) = \hat{\tilde{x}}_2(k) - \tilde{x}_2(k) \end{cases} \tag{7.5}$$

where $\hat{\tilde{x}}(k) = \left[\hat{\tilde{x}}_1(k)\,\hat{\tilde{x}}_2(k)\right]^T$. Let $\overline{A} = \begin{bmatrix} a_{11}\ a_{12} \\ a_{21}\ a_{22} \end{bmatrix}$, $\overline{B} = \begin{bmatrix} b_1 \\ b_2 \end{bmatrix}$, and a_{ij} and b_i are the corresponding system parameters. Hence, Eq. (7.1) can be rewritten as:

$$\begin{bmatrix} \tilde{x}_1(k+1) \\ \tilde{x}_2(k+1) \end{bmatrix} = \begin{bmatrix} a_{11}\ a_{12} \\ a_{21}\ a_{22} \end{bmatrix} \begin{bmatrix} \tilde{x}_1(k) \\ \tilde{x}_2(k) \end{bmatrix} + \begin{bmatrix} b_1 \\ b_2 \end{bmatrix} u(k) + \begin{bmatrix} \gamma_1(k) \\ \gamma_2(k) \end{bmatrix} \tag{7.6}$$

which equals:

$$\begin{cases} \tilde{x}_1(k+1) = a_{11}\tilde{x}_1(k) + a_{12}\tilde{x}_2(k) + b_1 u(k) + \gamma_1(k) \\ \tilde{x}_2(k+1) = a_{21}\tilde{x}_1(k) + a_{22}\tilde{x}_2(k) + b_2 u(k) + \gamma_2(k) \end{cases} \tag{7.7}$$

Based on the above, the STO can be designed as:

$$\begin{cases} \hat{\tilde{x}}_1(k+1) = a_{11}\hat{\tilde{x}}_1(k) + a_{12}\hat{\tilde{x}}_2(k) + b_1 u(k) + \vartheta_1(k) \\ \hat{\tilde{x}}_2(k+1) = a_{21}\hat{\tilde{x}}_1(k) + a_{22}\hat{\tilde{x}}_2(k) + b_2 u(k) + \vartheta_2(k) \\ \hat{\tilde{y}}_i(k) = \overline{C}_i \hat{\tilde{x}}(k) \end{cases} \tag{7.8}$$

where $\vartheta_1(k) = k_1|\tilde{e}_1(k)|^{\frac{1}{2}} sign(\tilde{e}_1(k)) + k_2\tilde{e}_1(k)$ and $\vartheta_2(k) = k_3 sign(\tilde{e}_1(k)) + k_4\tilde{e}_1(k)$ are the compensation terms.

Moreover, based on Eqs. (7.5), (7.7), and (7.8), the error of the states at $k+1$ can be formulated as:

$$\begin{cases} \tilde{e}_1(k+1) = a_{11}\tilde{e}_1(k) + a_{12}\tilde{e}_2(k) + \vartheta_1(k) - \gamma_1(k) \\ \tilde{e}_2(k+1) = a_{21}\tilde{e}_1(k) + a_{22}\tilde{e}_2(k) + \vartheta_2(k) - \gamma_2(k) \end{cases} \tag{7.9}$$

where $\gamma(k)$ is bounded, and k_1, k_2, k_3, and k_4 are the parameters of the STO. Eventually, both $\tilde{e}_1(k+1)$ and $\tilde{e}_2(k+1)$ will approach to zero, which is $\hat{\tilde{x}}(k) = \tilde{x}(k)$.

Furthermore, according to Eqs. (7.8) and (7.9), the error equation can be represented as:

$$\tilde{e}(k+1) = \overline{G}\tilde{e}(k) + \overline{H}(k)sign(\tilde{e}_1(k)) \tag{7.10}$$

where $\overline{G} = \begin{bmatrix} a_{11} + k_2 a_{12} \\ a_{21} + k_4 a_{22} \end{bmatrix}$, $\overline{H}(k) = \begin{bmatrix} k_1 |\tilde{e}_1(k)|^{\frac{1}{2}} \\ k_3 \end{bmatrix}$.

A lemma is given as follows to analyze the STO:

Lemma 1 [23]

$$\widehat{M}^T \widehat{N} + \widehat{N}^T \widehat{M} \le \widehat{M}^T \wp^{-1} \widehat{M} + \widehat{N}^T \wp \widehat{N} \tag{7.11}$$

Theorem 1 *For the given appropriate parameters* k_1, k_2, k_3, *and* k_4, *if there exist positive-definite matrices* $\widehat{X} = \widehat{X}^T > 0$, $\widehat{Q} = \widehat{Q}^T > 0$, *and* $\tilde{\wp} = \tilde{\wp}^T > 0$, *the following LMI holds*:

$$\overline{G}^T (\widehat{X} + \widehat{X}\tilde{\wp}\widehat{X})\overline{G} - (1 - \tilde{\aleph})\widehat{X} < -\widehat{Q} \tag{7.12}$$

Then the STO can be regarded as stable, and the trajectories of the observation error will converge into a ball formulated as $\widehat{Y}_r = \tilde{e} : \|\tilde{e}\|^2 < R_d$, the center of which is at the origin of the coordinates. $R_d = \frac{d}{1-\tilde{\aleph}}$ represents the radius of the ball.

Where $0 < \tilde{\aleph} < 1$, $d = \tilde{\overline{m}}_2 + \frac{1}{4}\tilde{\overline{m}}_1^2 \left\| \widehat{Q}^{-1} \right\|^2$, $\tilde{\overline{m}}_1 = \tilde{m}_1 + (k_1 k_3 f_{12})^2 \tilde{n}_1$, $\tilde{\overline{m}}_2 = k_3^2 f_{22} + (k_1 k_3 f_{12})^2 \tilde{n}_1^{-1}$, $\tilde{m}_1 = k_1^2 f_{11}$, $\tilde{n}_1 \in \Re^+$, $\overline{F} = \begin{bmatrix} f_{11}\ f_{12} \\ f_{21}\ f_{22} \end{bmatrix} = \begin{bmatrix} \tilde{\wp}^{-1} + \widehat{X} \end{bmatrix}$, $\overline{F}, \tilde{\wp}, \widehat{X} \in \Re^{2\times 2}$.

Proof We introduce the Lyapunov function as follows:

$$\overline{V}(k) = \tilde{e}^T(k)\widehat{X}\tilde{e}(k) \tag{7.13}$$

So that

$$\begin{aligned} \Delta\overline{V}(k+1) &= \overline{V}(k+1) - \overline{V}(k) \\ &= \tilde{e}^T(k+1)\widehat{X}\tilde{e}(k+1) - \tilde{e}^T(k)\widehat{X}\tilde{e}(k) \end{aligned} \tag{7.14}$$

Based on Eqs. (7.10), (7.14) can be written as:

$$\begin{aligned}\Delta\overline{V}(k+1) &= \tilde{e}^T(k)[\overline{G}^T\widehat{X}\overline{G} - \widehat{X}]\tilde{e}(k) + sign(\tilde{e}_1(k))\overline{H}^T(k)\widehat{X}\overline{G}\tilde{e}(k) \\ &\quad + \overline{H}^T(k)\widehat{X}\overline{H}(k) + \tilde{e}^T(k)\overline{G}^T\widehat{X}\overline{H}(k)sign(\tilde{e}_1(k)) \\ &= \tilde{e}^T(k)[\overline{G}^T\widehat{X}\overline{G} - \widehat{X}]\tilde{e}(k) + \overline{H}^T(k)\widehat{X}\overline{H}(k) + 2\tilde{e}^T(k)\overline{G}^T\widehat{X}\overline{H}(k)sign(\tilde{e}_1(k))\end{aligned} \tag{7.15}$$

According to Lemma 1, the term in Eq. (7.15) can be represented by an inequation as follows:

$$2\tilde{e}^T(k)\overline{G}^T\widehat{X}\overline{H}(k)sign(\tilde{e}_1(k)) \leq \tilde{e}^T(k)\overline{G}^T\widehat{X}\tilde{\wp}^{-1}\widehat{X}\overline{G}\tilde{e}(k) + \overline{H}^T(k)\tilde{\wp}\overline{H}(k) \tag{7.16}$$

Hence, Eq. (7.15) can be written as:

$$\begin{aligned}\Delta\overline{V}(k+1) &\leq \tilde{e}^T(k)[\overline{G}^T(\widehat{X} + \widehat{X}\tilde{\wp}\widehat{X})\overline{G} - (1-\tilde{\aleph})\widehat{X}]\tilde{e}(k) \\ &\quad + \overline{H}^T(k)[\tilde{\wp}^{-1} + \widehat{X}]\overline{H}(k) - \tilde{\wp}\overline{V}(k)\end{aligned} \tag{7.17}$$

Thus,

$$\begin{aligned}&\overline{H}^T(k)[\tilde{\wp}^{-1} + \widehat{X}]\overline{H}(k) \\ &\qquad = \overline{H}^T(k)\overline{F}\overline{H}(k) \\ &\qquad = \begin{bmatrix} k_1|\tilde{e}_1(k)|^{\frac{1}{2}} \\ k_3 \end{bmatrix}^T \begin{bmatrix} f_{11} & f_{12} \\ f_{21} & f_{22} \end{bmatrix} \begin{bmatrix} k_1|\tilde{e}_1(k)|^{\frac{1}{2}} \\ k_3 \end{bmatrix} \\ &\qquad = \tilde{m}_1|\tilde{e}_1(k)| + \tilde{m}_2|\tilde{e}_1(k)|^{\frac{1}{2}} + \tilde{m}_3\end{aligned} \tag{7.18}$$

where $\tilde{m}_1 = k_1^2 f_{11}$, $\tilde{m}_2 = k_1 k_3 f_{12}$, $\tilde{m}_2 = k_3^2 f_{22}$.

According to Lemma 1, the term in Eq. (7.18) can be further represented if there exists $\widehat{Q} = \widehat{Q}^T > 0$, which satisfies the equation as follows:

$$\overline{G}^T(\widehat{X} + \widehat{X}\tilde{\wp}\widehat{X})\overline{G} - (1-\tilde{\aleph})\widehat{X} + \widehat{Q} = 0 \tag{7.19}$$

Then $\Delta\overline{V}(k)$ can be written as:

$$\Delta\overline{V}(k+1) \leq -\|\tilde{e}(k)\|_Q^2 + \tilde{\overline{m}}_1|\tilde{e}_1(k)| - \tilde{\aleph}\overline{V}(k) + \tilde{\overline{m}}_2 \tag{7.20}$$

where $\tilde{\overline{m}}_1 = \tilde{m}_1 + (k_1 k_3 f_{12})^2 \tilde{n}_1$, $\tilde{\overline{m}}_2 = k_3^2 f_{22} + (k_1 k_3 f_{12})^2 \tilde{n}_1^{-1}$.

Moreover, Eq. (7.20) can be derived as follows:

$$\Delta\overline{V}(k+1) \leq -\|\tilde{e}(k)\|^2_{\widehat{Q}} + \tilde{\overline{m}}_1 \left\| \widehat{Q}\widehat{Q}^{-1} \tilde{e}_1(k) \right\| - \tilde{\aleph}\overline{V}(k) + \tilde{\overline{m}}_2 \tag{7.21}$$

$$\begin{aligned}\Delta\overline{V}(k+1) \leq & -\left[\left\| \widehat{Q}^{\frac{1}{2}} \tilde{e}(k) \right\| - \frac{1}{2}\tilde{\overline{m}}_1 \left\| \widehat{Q}^{-1} \right\| \right]^T \times \left[\left\| \widehat{Q}^{\frac{1}{2}} \tilde{e}(k) \right\| - \frac{1}{2}\tilde{\overline{m}}_1 \left\| \widehat{Q}^{-1} \right\| \right] \\ & + \frac{1}{4}\tilde{\overline{m}}_1^2 \left\| \widehat{Q}^{-1} \right\|^2 - \tilde{\aleph}\overline{V}(k) + \tilde{\overline{m}}_2 \end{aligned} \tag{7.22}$$

That is,

$$\Delta\overline{V}(k+1) \leq -\tilde{\aleph}\overline{V}(k) + d \tag{7.23}$$

Hence,

$$\overline{V}(k+1) \leq (1-\tilde{\aleph})\overline{V}(k) + d \tag{7.24}$$

By iterating Eq. (7.24), the following conclusion can be obtained:

$$\overline{V}(k+1) \leq (1-\tilde{\aleph})^k \overline{V}(0) + \sum_{j=0}^{k} (1-\tilde{\aleph})^{j-1} d \tag{7.25}$$

Inequation (7.25) indicates that the observation error is convergent. Furthermore, after finite steps k, $\lim_{k \to N} \overline{V}(k) \leq \frac{d}{1-\tilde{\aleph}}$, and N is a finite number. $R_d \leq \frac{d}{1-\tilde{\aleph}}$, which means the trajectories of the observation errors have converged into a ball. This finishes the proof.

7.3.2 Design of the Super-Twisting Controller

We define the output speed of the system as follows:

$$\sigma(k) = \tilde{\zeta}(k) - \tilde{y}_r(k) \tag{7.26}$$

where $\zeta(k)$ is the actual output speed of the DC motor system, and $y_r(k)$ is the reference input speed. The discrete-time sliding mode function is designed as follows:

$$\tilde{s}(k) = \lambda_1\sigma(k) + \lambda_2 \sum_{\tau=t_1(k)}^{t_2(k)} sig^{\partial}(\sigma(\tau)) + \lambda_3 \vartriangle \sigma(k) \tag{7.27}$$

where $\begin{cases} t_1(k) = \max\{k - 10, 0\} \\ t_2(k) = \max\{k - 1, 0\} \end{cases}$, $\lambda_1 > 0$, $\lambda_2 > 0$, $\lambda_3 > 0$, $0 < \partial < 1$ are the gain parameters, and $sig^{\partial}(\kappa)$, $\triangle\ \varepsilon(k)$ are defined by $sig^{\partial}(\kappa) \triangleq |\kappa|^{\partial} sign(\kappa)$, $\triangle\ \sigma(k) \triangleq \sigma(k) - \sigma(k-1)$.

Remark 2 The parameters $t_1(k)$, $t_2(k)$ in Eq. (7.27) are variables that limit the data length of the sum term in Eq. (7.27) and ensure that only the recent ten data points are summed (if they exist), reducing the instability caused by residual accumulation. In short, let $t_1 = t_1(k)$, $t_2 = t_2(k)$.

Assumption 2 The error value of the reference input and actual output $\varepsilon(i)= 0i \leq 0$, and the control input $u(i)= 0i \leq 0$.

Thus, $\tilde{s}(k{+}1)$ can be formulated as follows:

$$\tilde{s}(k{+}1) = \lambda_1\sigma(k+1) + \lambda_2 \sum_{\tau=t_1}^{t_2} sig^{\partial}(\sigma(\tau)) + \lambda_3\Delta\sigma(k+1) \tag{7.28}$$

When the sliding mode is in equilibrium, $\tilde{s}(k{+}1) = 0$, which is:

$$\lambda_1\sigma(k+1) + \lambda_2 \sum_{\tau=t_1}^{t_2} sig^{\partial}(\sigma(\tau)) + \lambda_3\Delta\sigma(k+1) = 0 \tag{7.29}$$

Further, we substitute Eqs. (7.26) and (7.27) into Eq. (7.29) and derive:

$$\begin{aligned}
\tilde{s}(k{+}1) &= \lambda_1\sigma(k+1) + \lambda_2 \sum_{\tau=t_1}^{t_2} sig^{\partial}(\sigma(\tau)) + \lambda_3(\sigma(k+1) - \sigma(k)) \\
&= (\lambda_1 + \lambda_3)\sigma(k+1) + \lambda_2 \sum_{\tau=t_1}^{t_2} sig^{\partial}(\sigma(\tau)) - \lambda_3\sigma(k) \\
&= (\lambda_1 + \lambda_3)(\tilde{\zeta}(k+1) - \tilde{y}_r(k+1)) \\
&\quad + \lambda_2 \sum_{\tau=t_1}^{t_2} sig^{\partial}(\sigma(\tau)) - \lambda_3\sigma(k) \\
&= (\lambda_1 + \lambda_3)(\overline{C}_i(\overline{A}\hat{\tilde{x}}(k) + \overline{B}u_{eq}(k)) - \overline{y}_r(k+1)) \\
&\quad + \lambda_2 \sum_{\tau=t_1}^{t_2} sig^{\partial}(\sigma(\tau)) - \lambda_3\sigma(k) \\
&= 0
\end{aligned} \tag{7.30}$$

Thus, the equivalent control input equation is:

$$u_{eq}(k) = \overline{B}^{-1}\left[\overline{C}_i^{-1}\left(\frac{-\lambda_2 \sum_{\tau=t_1}^{t_2} sig^{\partial}(\sigma(\tau)) + \lambda_3 \sigma(k)}{(\lambda_1 + \lambda_3)} + \tilde{y}_r(k+1)\right) - \overline{A}\hat{\tilde{x}}(k)\right] \tag{7.31}$$

The equivalent control law has the following form:

$$u(k) = u_{eq}(k) + u_s(k) \tag{7.32}$$

where $u_s(k)$ is the control input in the form of the STA.

Hence, $u_s(k)$ can be designed as follows:

$$u_s(k) = -\overline{\Gamma}_1 |\tilde{s}(k)|^{\frac{1}{2}} sign(\tilde{s}(k)) + \tilde{z}(k) \tag{7.33}$$

$$\tilde{z}(k+1) = -\overline{\Gamma}_2 sign(\tilde{s}(k)) \tag{7.34}$$

where $\overline{\Gamma}_1 > 0$ and $\overline{\Gamma}_2 > 0$ are the constant gains, which need to be designed.

Remark 3 Due to $u_s(k)$ has the form of an STA, demonstrating that the form is mathematically stable. The proposed SSTSMC is based on a second-order SMC law, which removes the discontinuous switching function from first order control law and hides it in higher order term. In this way, the jitter caused by the switching function is effectively suppressed.

7.3.3 Analysis of Sliding Motion

Definition 1 If the following inequation is satisfied, then sliding motion accessibility is proved [22].

$$|\tilde{s}(k+1)| \leq |\tilde{s}(k)| \tag{7.35}$$

Theorem 2 If the abovementioned inequation is satisfied, for the discrete-time system in (7.1), the sliding mode function in Eq. (7.27), and the equivalent control law in Eq. (7.32), the trajectories of the system reach the sliding mode bandwidth within finite steps.

Proof The proof process can be separated into two steps, i.e., sliding motion accessibility and convergence in finite steps.

$$\begin{aligned}\sigma(k+1) &= \tilde{\zeta}(k+1) - \tilde{y}_r(k+1) \\ &= \overline{C}_i \hat{\tilde{x}}(k+1) - \tilde{y}_r(k+1) \\ &= \overline{C}_i \overline{A} \hat{\tilde{x}}(k) + \overline{C} \overline{B} u(k) - \tilde{y}_r(k+1)\end{aligned} \tag{7.36}$$

According to Eqs. (7.31) and (7.32), Eq. (7.36) can be rewritten as:

$$\sigma(k+1) = \frac{-\lambda_2 \sum_{\tau=t_1}^{t_2} sig^{\partial}(\sigma(\tau)) + \lambda_3 \sigma(k)}{(\lambda_1 + \lambda_3)} + \overline{C}_i \overline{B} u_s(k) \tag{7.37}$$

Furthermore,

$$\begin{aligned}\Delta\sigma(k+1) &= \sigma(k+1) - \sigma(k) \\ &= \frac{-\lambda_2 \sum_{\tau=t_1}^{t_2} sig^{\partial}(\sigma(\tau)) + \lambda_3 \sigma(k)}{(\lambda_1 + \lambda_3)} + \overline{C}_i \overline{B} u_s(k) \\ &\quad - \frac{-\lambda_2 \sum_{\tau=t_1}^{t_2} sig^{\partial}(\sigma(\tau)) + \lambda_3 \sigma(k-1)}{(\lambda_1 + \lambda_3)} - \overline{C}_i \overline{B} u_s(k-1) \\ &= \frac{-\lambda_2 sig^{\partial}(\sigma(k)) + \lambda_3 \Delta\sigma(k)}{(\lambda_1 + \lambda_3)} + \overline{C}_i \overline{B} \Delta u_s(k)\end{aligned} \tag{7.38}$$

Case 1: When $\tilde{s}(k) \geq 0$

$$\begin{aligned}\tilde{s}(k+1) - \tilde{s}(k) &= \lambda_1 \sigma(k+1) + \lambda_2 \sum_{\tau=t_1}^{t_2} sig^{\partial}(\sigma(\tau)) \\ &\quad + \lambda_3 \Delta\sigma(k+1) - \lambda_1 \sigma(k) \\ &\quad - \lambda_2 \sum_{\tau=t_1}^{t_2} sig^{\partial}(\sigma(\tau)) - \lambda_3 \Delta\sigma(k) \\ &= \lambda_1 \Delta\sigma(k+1) + \lambda_2 sig^{\partial}(\sigma(k)) + \lambda_3(\Delta\sigma(k+1) - \Delta\sigma(k)) \\ &= (\lambda_1 + \lambda_3)\Delta\sigma(k+1) + \lambda_2 sig^{\partial}(\sigma(k)) - \lambda_3 \Delta\sigma(k)\end{aligned} \tag{7.39}$$

We substitute $\Delta\sigma(k+1)$ into Eq. (7.39) and rewrite it as:

$$\tilde{s}(k+1) - \tilde{s}(k) = (\lambda_1 + \lambda_3)\overline{C B} \Delta u_s(k) \tag{7.40}$$

$$
\begin{aligned}
\Delta u_s(k) &= u_s(k) - u_s(k-1) \\
&= \overline{\Gamma}_1|\tilde{s}(k)|^{\frac{1}{2}} sign(\tilde{s}(k)) + \tilde{z}(k) - \overline{\Gamma}_1|\tilde{s}(k-1)|^{\frac{1}{2}} sign(\tilde{s}(k-1)) - \tilde{z}(k-1) \\
&= \overline{\Gamma}_1\left[\tilde{s}(k)^{\frac{1}{2}} - \tilde{s}(k-1)^{\frac{1}{2}}\right] + \overline{\Gamma}_2 sign(\tilde{s}(k-1)) - \overline{\Gamma}_2 sign(\tilde{s}(k-2)) \\
&= \overline{\Gamma}_1\left[\tilde{s}(k)^{\frac{1}{2}} - \tilde{s}(k-1)^{\frac{1}{2}}\right]
\end{aligned}
\tag{7.41}
$$

Hence,

$$
\tilde{s}(k+1) - \tilde{s}(k) = (\lambda_1 + \lambda_3)\overline{C}_i\overline{B\Gamma}_1\left[\tilde{s}(k)^{\frac{1}{2}} - \tilde{s}(k-1)^{\frac{1}{2}}\right] \tag{7.42}
$$

where $\lambda_1, \lambda_3, \overline{\Gamma}_1 \geq 0$, $\overline{C}_i\overline{B}$ is the parameter of the plant, $(\lambda_1 + \lambda_3)\overline{C}_i\overline{B\Gamma}_1 \geq 0$.

Thus, if $\tilde{s}(k) - \tilde{s}(k-1) \leq 0$, $\tilde{s}(k+1) - \tilde{s}(k) \leq 0$ can be substituted.

$$
\tilde{s}(1) - \tilde{s}(0) = \lambda_1(\sigma(1) - \sigma(0)) + \lambda_2 sig^{\partial}(\sigma(0)) + \lambda_3(\sigma(1) - 2\sigma(0) + \sigma(-1)) \tag{7.43}
$$

According to Assumption 2, $\sigma(0) = \sigma(-1) = 0$ so that

$$
\begin{aligned}
\tilde{s}(1) - \tilde{s}(0) &= \lambda_1\sigma(1) + \lambda_3\sigma(1) \\
&= (\lambda_1 + \lambda_3)\cdot(\tilde{\zeta}(1) - \tilde{y}_r(1)) \\
&= (\lambda_1 + \lambda_3)\cdot(\overline{C}_i(\overline{A}\hat{\tilde{x}}(0) + \overline{B}u(0) - \tilde{y}_r(1))
\end{aligned}
\tag{7.44}
$$

where $\hat{\tilde{x}}(0) = \left[\hat{\tilde{x}}_1(0)\,\hat{\tilde{x}}_2(0)\right]^T = u(0) = 0, \lambda_1, \lambda_3, \tilde{y}_r(1) \geq 0$ so that $\tilde{s}(1) - \tilde{s}(0) \leq 0$.

We substitute $\tilde{s}(1) - \tilde{s}(0) \leq 0$ into Eq. (7.44) so that $\tilde{s}(2) - \tilde{s}(1) \leq 0, \tilde{s}(3) - \tilde{s}(2) \leq 0, \ldots, \tilde{s}(k+1) - \tilde{s}(k) \leq 0$ can be proved.

Case 2: When $\tilde{s}(k) \leq 0$, it can be concluded similarly that $\tilde{s}(k+1) - \tilde{s}(k) \geq 0$.

Combined with the above two cases, it has been demonstrated that

$$
|\tilde{s}(k+1)| \leq |\tilde{s}(k)| \tag{7.45}
$$

According to Definition 1, this finishes the proof of the sliding motion accessibility.

If we assume $\tilde{s}(k) \geq 0$, as proved above $\tilde{s}(1) - \tilde{s}(0) \leq 0$; thus, $\tilde{s}(1) - \tilde{s}(0) \leq \tilde{s}(1)^{\frac{1}{2}} - \tilde{s}(0)^{\frac{1}{2}} \leq 0$. We define $\tilde{s}(1) - \tilde{s}(0) = \tilde{\varpi} \leq 0$, $(\lambda_1 + \lambda_3)\overline{C}_i\overline{B\Gamma}_1 = \tilde{\delta}$ so that

$$
\tilde{s}(2) - \tilde{s}(1) = \tilde{\delta}\left[\tilde{s}(1)^{\frac{1}{2}} - \tilde{s}(0)^{\frac{1}{2}}\right] \geq \tilde{\delta}\tilde{\varpi} \tag{7.46}
$$

After iteration, the following is obtained

$$\tilde{s}(k+1) - \tilde{s}(k) = \tilde{\delta}\Big[\tilde{s}(k)^{\frac{1}{2}} - \tilde{s}(k-1)^{\frac{1}{2}}\Big] \geq \tilde{\delta}^k \tilde{\varpi} \tag{7.47}$$

and $\tilde{s}(k+1) - \tilde{s}(k) \leq 0$ so that

$$\tilde{\delta}^k \tilde{\varpi} \leq \tilde{s}(k+1) - \tilde{s}(k) \leq 0 \tag{7.48}$$

We use the step $\tau > 0$; after finite steps k, the sliding motion will reach the equilibrium state and the convergence time $T \leq \left|\tilde{\delta}^k \tilde{\varpi}\right| / \tau$. If $\tilde{s}(k) \leq 0$, it can be concluded similarly that the convergence time $T \leq \left|\tilde{\delta}^k \tilde{\varpi}\right| / \tau$. This finishes the proof.

7.4 Simulation Results

A simulation of the DC motor system is performed to demonstrate the effectiveness of the proposed method.

The parameters of the DC motor system are listed as follows:

$$\overline{A} = \begin{bmatrix} 0.9647 & -0.0282 \\ 0.0196 & 0.9997 \end{bmatrix}, \overline{B} = \begin{bmatrix} 0.0196 \\ 0.0002 \end{bmatrix}, \overline{C}_0 = \begin{bmatrix} -0.2293 \\ 1.4393 \end{bmatrix}^T,$$

$$\overline{C}_1 = \begin{bmatrix} -0.2 \\ 1.45 \end{bmatrix}^T, \overline{C}_2 = \begin{bmatrix} -0.25 \\ 1.5 \end{bmatrix}^T.$$

External disturbance $\gamma(k) = \begin{bmatrix} \gamma_1(k) \\ \gamma_2(k) \end{bmatrix}$, $w_1(k) = 3rand(k) + \sin(k)$, $w_2(k) = rand(k) + \cos(2k)$,$rand(k) \in [0, 1]$. The parameters of the STO are set as $k_1 = -5, k_2 = -0.6, k_3 = 1, k_4 = 0.2$. The sliding mode function: $\lambda_1 = 900$, $\lambda_2 = 150$, $\lambda_3 = 100$, $\partial = 0.3$, $\Gamma_1 = 65$, $\Gamma_2 = 8000$. The transfer matrix of the Markov jump process is $\pi(k) = \begin{bmatrix} 0.3\,0.4\,0.3 \\ 0.7\,0.2\,0.1 \\ 0.6\,0.2\,0.2 \end{bmatrix}$.

The simulation time is 30 s, and the sampling period is 0.02 s, resulting in 1500 sampling points. The initial conditions are as follows: $\tilde{x}(0) = [1000\ 500]^T$, $\hat{\tilde{x}}(0) = [0\ 0]^T$, $u(0) = 0$, $\tilde{y}_r(0) > 0$.

Figure 7.2 shows the external disturbances injected into the DC motor system. The disturbances are hybrid sinusoidal functions with different frequencies and amplitudes. The disturbances are used to simulate interferences in a real environment. The two components are injected into the systems in two states.

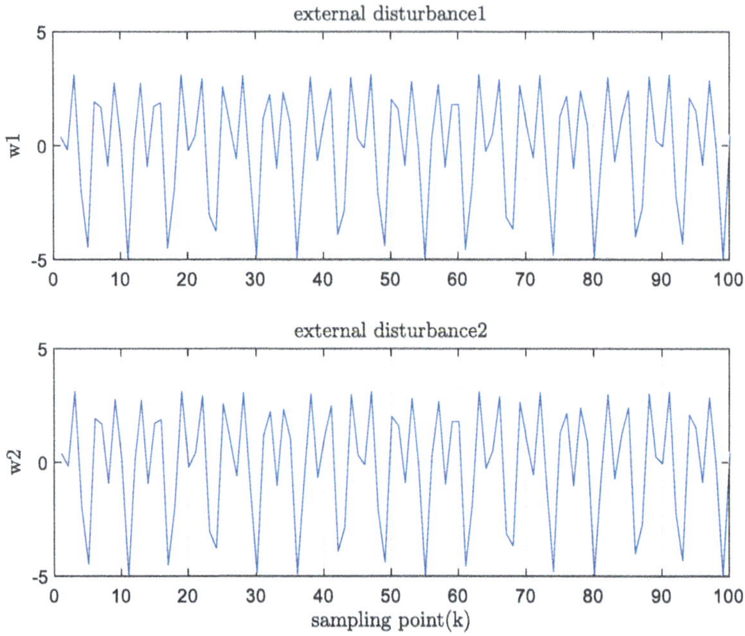

Fig. 7.2 External disturbances injected into the DC motor system

The tracking trajectories of the DC motor system at the speed of 2000 r/min are shown in Fig. 7.3. The proposed SSTSMC reacts quicker than the compared 2-SMC. When the system starts to work, stabilization occurs at about the 85th sampling point for the SSTSMC and at 80 sampling points later for the 2-SMC. When the system is stable, the proposed method results in a lower static error and less jitter than the 2-SMC. Moreover, when the channels changes, both methods exhibit jitter. It is worth noting that the proposed method can retrace the trajectory rapidly with little jitter in this case.

Figure 7.4 shows the control input of the DC motor system at the speed of 2000 r/min. As the channel changes, the proposed method consumes less energy for restabilization. When the system is stable, the SSTSMC produces little jitter.

In addition, simulations of the DC motor system are performed at speeds ranging from 1000 to 2500 r/min to simulate the working state of the motor in a real production environment. The results are shown in Figs. 7.5 and 7.6.

Figure 7.5 shows the tracking trajectories of the DC motor system at speeds varying from 1000 to 2500 r/min. In the first, third, and final stage, the speed remains constant, and the system controlled by the proposed method reacts quicker than that controlled by the 2-SMC, which exhibits considerable jitter. In the second and fourth stages, the speed changes rapidly from 1000 to 2000 r/min and from 2500 to 2000 r/min. The trajectories of both methods are similar, but the SSTSMC generates little jitter. The control input of the DC motor system at variable speed is shown in Fig. 7.6.

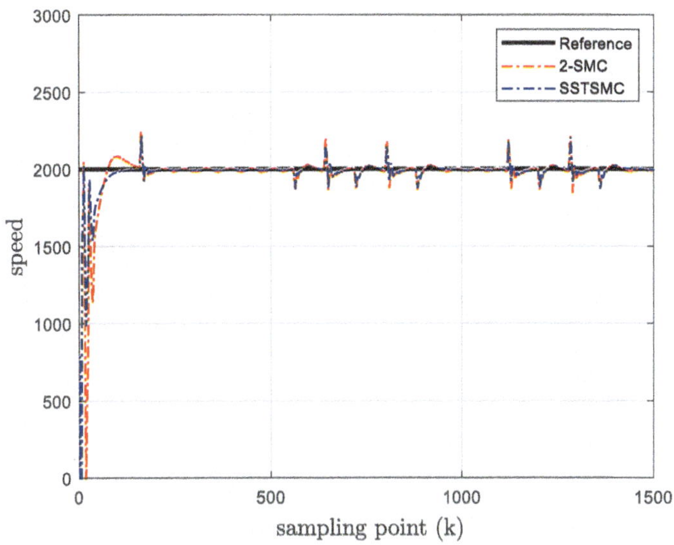

Fig. 7.3 Tracking output of the DC motor system at fixed speed

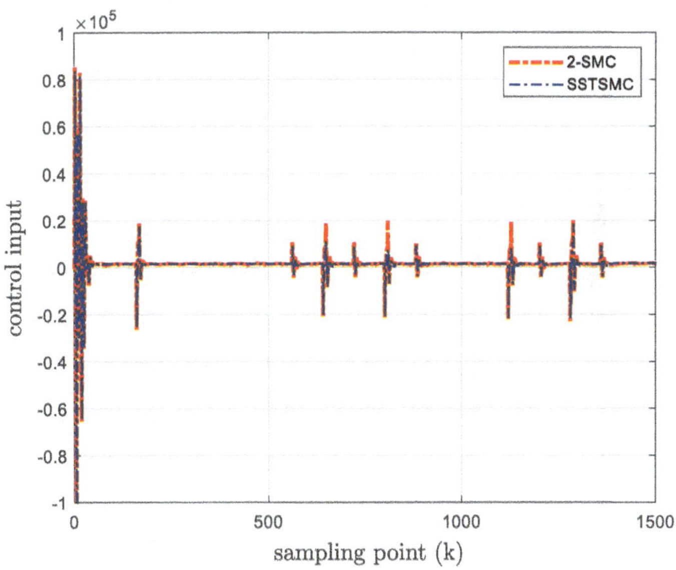

Fig. 7.4 Control input of the DC motor system at fixed speed

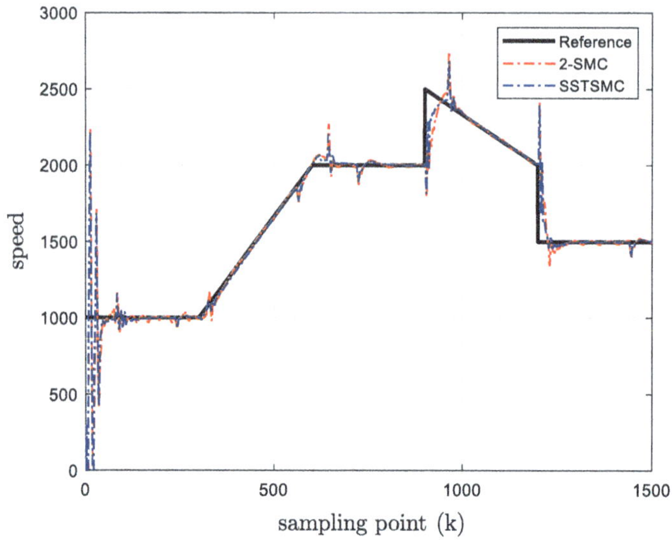

Fig. 7.5 Tracking output of the DC motor system at changing speed

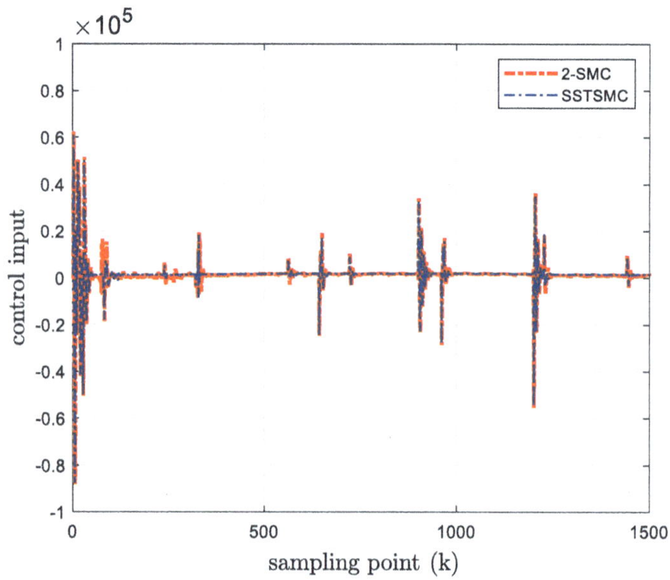

Fig. 7.6 Control input of the DC motor system at changing speed

The SSTSMC generates less jitter and consumes less energy than the SSTSMC, and produces a stable response.

7.5 Conclusions

In this chapter, a DC motor system with switching channels and external disturbances was analyzed. Instability and jitter were introduced. A DC motor model based on Markov jump processes was established to solve these problems. An STO was used to estimate the states and suppress the jitter. Furthermore, an SSTSMC based on the STA was designed to improve the control performance. The reachability of the sliding surface, the convergence in finite steps, and the stability of the controller were mathematically analyzed. Numerical simulations were conducted. The results showed that the proposed SSTSMC method was more reliable and provided better control performance than the conventional 2-SMC strategy.

References

1. A. Attar, J. Bouchnaif, K. Grari, "Control of Brushless DC motors using sensorless Back-EMF integration method," Materials Today: Proceedings,2021.
2. L. Zhang, X. Yin, Z. Ning and D. Ye, "Robust filtering for a class of networked nonlinear systems with switching communication channels," IEEE Transactions on Cybernetics, vol. 47, no. 3, pp. 671-682, Mar. 2017.
3. Z. Wu, P. Shi, Z. Shu, H. Su and R. Lu, "Passivity-based asynchronous control for Markov jump systems," IEEE Transactions on Automatic Control, vol. 62, no. 4, pp. 2020-2025, Apr. 2017.
4. X. Yin, L. Zhang, Y. Zhu, C. Wang and Z. Li, "Robust control of networked systems with variable communication capabilities and application to a semi-active suspension system," IEEE/ASME Transactions on Mechatronics, vol. 21, no. 4, pp. 2097-2107, Aug. 2016.
5. J. Zhu, X. Yu, T. Zhang, et al. "Sliding mode control of MIMO Markovian jump systems." Automatica, vol. 68, pp. 286-293, Jun. 2016.
6. H. Youness, M. Moness and M. Khaled, "MPSoCs and multicore microcontrollers for embedded PID control: a detailed study," IEEE Transactions on Industrial Informatics, vol. 10, no. 4, pp. 2122-2134, Nov. 2014.
7. G. Rigatos, P. Siano, A. Melkikh and N. Zervos, "A nonlinear H-infinity control approach to stabilization of distributed synchronous generators," IEEE Systems Journal, vol. 12, no. 3, pp. 2654-2663, Sept. 2018.
8. Z. Du, D. Yue and S. Hu, "H-infinity stabilization for singular networked cascade control systems with state delay and disturbance," IEEE Transactions on Industrial Informatics, vol. 10, no. 2, pp. 882-894, May 2014.
9. H. Sun, R. Yu, Y. Chen and H. Zhao, "Optimal design of robust control for fuzzy mechanical systems: performance-based leakage and confidence-index measure," IEEE Transactions on Fuzzy Systems, vol. 27, no. 7, pp. 1441-1455, Jul. 2019.
10. H. Yan, Y. Xu, F. Cai, H. Zhang, W. Zhao and C. Gerada, "PWM-VSI Fault Diagnosis for a PMSM Drive Based on the Fuzzy Logic Approach," IEEE Transactions on Power Electronics, vol. 34, no. 1, pp. 759-768, Jan. 2019.

11. G. P. Incremona, A. Ferrara and L. Magni, "Asynchronous networked MPC with ISM for uncertain nonlinear systems," IEEE Transactions on Automatic Control, vol. 62, no. 9, pp. 4305-4317, Sept. 2017.
12. M. Zanon, L. Grüne and M. Diehl, "Periodic Optimal Control, Dissipativity and MPC," IEEE Transactions on Automatic Control, vol. 62, no. 6, pp. 2943-2949, Jun. 2017.
13. R. Han, L. Meng, J. M. Guerrero and J. C. Vasquez, "Distributed Nonlinear Control with Event-Triggered Communication to Achieve Current-Sharing and Voltage Regulation in DC Microgrids," IEEE Transactions on Power Electronics, vol. 33, no. 7, pp. 6416-6433, July 2018.
14. B. Demirel, A. S. Leong, V. Gupta and D. E. Quevedo, "Tradeoffs in Stochastic Event-Triggered Control," IEEE Transactions on Automatic Control, vol. 64, no. 6, pp. 2567-2574, Jun. 2019.
15. S. Zou, A. Mallik, J. Lu and A. Khaligh, "Sliding Mode Control Scheme for a CLLC Resonant Converter," IEEE Transactions on Power Electronics, vol. 34, no. 12, pp. 12274-12284, Dec. 2019.
16. Q. Xu, "Piezoelectric nanopositioning control using second-order discrete-time terminal sliding-mode strategy," IEEE Transactions on Industrial Electronics, vol. 62, no. 12, pp. 7738-7748, Dec. 2015.
17. A. Ferrara and G. P. Incremona, "Design of an integral suboptimal second-order sliding mode controller for the robust motion control of robot manipulators," IEEE Transactions on Control Systems Technology, vol. 23, no. 6, pp. 2316-2325, Nov. 2015.
18. A. Merabet, K. T. Ahmed, H. Ibrahim and R. Beguenane, "Implementation of sliding mode control system for generator and grid sides control of wind energy conversion system," IEEE Transactions on Sustainable Energy, vol. 7, no. 3, pp. 1327-1335, July 2016.
19. Q. Xu, "Continuous integral terminal third-order sliding mode motion control for piezoelectric nanopositioning System," IEEE/ASME Transactions on Mechatronics, vol. 22, no. 4, pp. 1828-1838, Aug, 2017.
20. J. A. Moreno and M. Osorio, "Strict lyapunov functions for the super-twisting algorithm," IEEE Transactions on Automatic Control, vol. 57, no. 4, pp. 1035-1040, Apr, 2012.
21. L. Zou, Z. Wang, H. Gao, "Observer-based H∞ control of networked systems with stochastic communication protocol: The finite-horizon case," Automatica, vol. 63, pp. 366-373, Jan. 2016.
22. M. Li and Y. Chen, "Robust adaptive sliding mode control for switched networked control systems with disturbance and faults," IEEE Transactions on Industrial Informatics, vol. 15, no. 1, pp. 193-204, Jan. 2019.
23. L. Qiu, Y. Shi, F. Yao, G. Xu, and B. Xu, "Network-based robust H2/H∞ control for linear systems with two-channel random packet dropouts and time delays," IEEE Transactions on Cybernetics, vol. 45, no. 8, pp. 1450-1462, Aug. 2015.

Chapter 8
Resilient Sliding-Mode Tracking Control for Networked Systems with DoS Attack

This chapter investigates the speed tracking control of multi-sensor servo systems with external interference and denial of service (DoS) attacks in network control systems (NCSs) in recent years. First, an observer based on the super-twisting algorithm is proposed to observe and estimate external interference. Second, a weight-based try-once-discard (TOD) protocol is used to select multiple sensors for network transmission, and the most critical data are selected for transmission. Then, we model DoS attacks in network control and implement network compensation measures to reduce the possibility of system instability. On this basis, a sliding mode control (SMC) algorithm is designed to implement the closed-loop control of the system. Experiments and simulations prove the validity and feasibility of the proposed approach.

8.1 Introduction

Due to the rapid development of computer and automation control technology, NCSs are extensively researched. An NCS is a remote control system whose controllers, actuators, and sensors may be distributed in many areas, and the signals are transmitted through the network [1]. The emergence of the NCS has simplified and improved control technology. However, there are problems caused by the uncertainty of network switching, such as network attacks and security breaches, network congestion and delay, network load changes, and distribution problems.

Many scholars investigated cyber-attacks. In [2], Li et al. summarized the problems of NCSs under a false data injection (FDI) attack, which was considered a game process in which the attacker selected the weakest part of the system as the intrusion point. The NCS side needs to allocate defense resources to maintain maximum data protection. In [3], a new optimal control method was proposed to prevent network-induced dropouts. A novel Smith packet loss predictor was designed to predict the

© The Author(s), under exclusive license to Springer Nature Singapore Pte Ltd. 2022

M. Li et al., *Tracking Control of Networked Systems via Sliding-Mode*,
https://doi.org/10.1007/978-981-16-6514-1_8

current state based on historical data of the communication network to compensate for losses without any knowledge of the system dynamics.

Different protocol models were abstracted from communication network constraints to solve network uncertainty problems under different conditions. The round-robin (RR) protocol, TOD protocol, and stochastic communication (SC) protocol are the three main communication protocols. In communication transmission, only one data node of the network is used for transmission. These protocols have specific rules for selecting nodes. In [4], moving horizon estimation of a class of discrete time-delay systems was conducted using the RR protocol. The nodes were selected sequentially. In [5], the TOD protocol was used to prevent data conflicts in linear systems with multiple sensors. The protocol was updated and improved so that the sensor with the largest error sent data first.

Various advanced control algorithms have been proposed, including proportional–integral–derivative (PID) control [6], fuzzy control [7, 8], H-infinity control [5, 9, 10], neural network control [11, 12], event trigger control [13, 14], and SMC [15, 16]. SMC has the advantages of a simple algorithm, strong robustness, and fast response speed. It prevents uncertainty of the system and is suitable for nonlinear systems. Some scholars applied the SMC method to space robot control. SMC is an important nonlinear control method and has been extensively studied in recent years. However, there are several shortcomings of SMC. When the SMC's motion trajectory is close to the sliding surface, it usually moves back and forth within the sliding mode bandwidth instead of directly moving to the equilibrium point, representing a considerable problem in practical applications.

Numerous studies investigated methods to suppress jitter, including high-order SMC and adaptive SMC. Xu [16] proposed an SMC strategy based on a terminal sliding surface and designed a new second-order discrete signal for the precise motion control strategy of an SMC-based piezoelectric-driven nano-positioning equipment. By using the feedback of the output value, this strategy was easy to implement and did not require the use of a state observer. In [17], an integral suboptimal second-order SMC algorithm was proposed. The controller adjusted by this algorithm was different from the conventional SMC, which provides continuous control to the system; thus, it is more suitable for the operation of industrial robots. The algorithm has a positive damping effect and was demonstrated for use in the industry. Merabet et al. [18] designed an SMC method to control the wind energy conversion between the generator and the grid side. This method is based on the external interference model and has strong robustness. The experiments showed that the method was robust to uncertain interference, parameter changes, and system uncertainty.

Inspired by these papers, we construct a multi-sensor servo system with external interference and DoS attacks. The main structure is shown in Fig. 8.1. The selection of multiple sensors is determined by the weight-based TOD protocol. A super-twisting observer (STO) is proposed to eliminate system state errors caused by switching sensors and external interference. In addition, DoS attacks are considered, and corresponding network compensation strategies are designed to reduce the possibility of system instability. In addition, an SMC method is proposed to control the servo

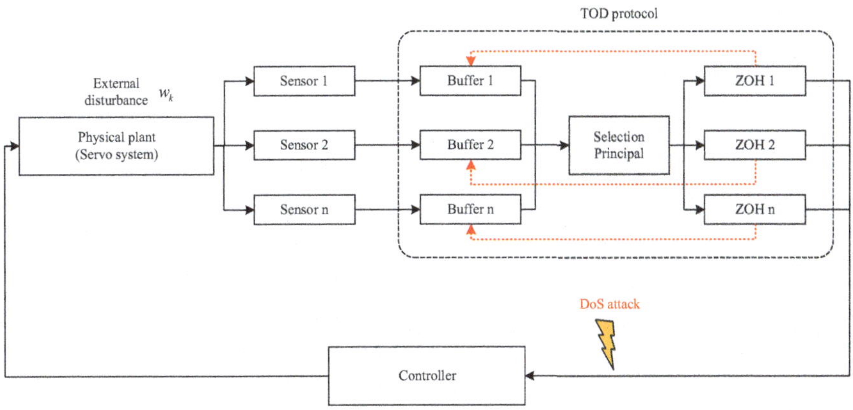

Fig. 8.1 Multi-sensor servo systems with external interference under DoS attack

system. Finally, the proposed control method is analyzed using numerical simulations and experiments. The results demonstrate the effectiveness of the proposed method.

The rest of this chapter is organized as follows. In Sect. 8.2, the external disturbance of the servo system is modeled using the STO to reduce disturbances. In Sect. 8.3, the sensor selection according to the weight-based TOD protocol is described. DoS attacks are analyzed, and compensation for data loss caused by the network is performed. An SMC is designed. Section 8.4 presents the simulation and experimental results. Section 8.5 concludes the chapter.

8.2 External Disturbance and Observer

The servo system with external disturbance is investigated, and a second-order observer with a super-twisting algorithm is proposed to damp out the external disturbance.

We consider the following two-dimensional system:

$$\overline{x}(k+1) = \overline{G}\overline{x}(k) + \overline{H}u(k) + \overline{w}(k) \tag{8.1}$$

where $\overline{x}(k) = [\overline{x}_1(k), \overline{x}_2(k)]^T$ and $u(k)$ are the state and control input, respectively. $\overline{w}(k) = [\overline{w}_1(k), \overline{w}_2(k)]^T$ is the disturbance introduced by the network or system parameter drift. $\overline{G}$ and $\overline{H}$ are the system matrices.

Assumption 1 We assume that the external disturbance $\overline{w}(k)$ is bounded, and the system pair $\left(\overline{G}, \overline{H}\right)$ is controllable for simplification.

The errors of the system states are defined as follows:

$$\overline{e}_1(k) = \hat{\overline{x}}_1(k) - \overline{x}_1(k) \tag{8.2}$$

$$\overline{e}_2(k) = \hat{\overline{x}}_2(k) - \overline{x}_2(k) \tag{8.3}$$

where $\hat{\overline{x}}(k) = \begin{bmatrix} \hat{\overline{x}}_1(k) \\ \hat{\overline{x}}_2(k) \end{bmatrix}$. Let $\overline{G} = \begin{bmatrix} \overline{a}_{11}\,\overline{a}_{12} \\ \overline{a}_{21}\,\overline{a}_{22} \end{bmatrix}$, $\overline{H} = \begin{bmatrix} \overline{b}_1 \\ \overline{b}_2 \end{bmatrix}$, hence, Eq. (8.1) can be represented as:

$$\begin{bmatrix} \overline{x}_1(k+1) \\ \overline{x}_2(k+1) \end{bmatrix} = \begin{bmatrix} \overline{a}_{11}\ \overline{a}_{12} \\ \overline{a}_{21}\ \overline{a}_{22} \end{bmatrix} \begin{bmatrix} \overline{x}_1(k) \\ \overline{x}_2(k) \end{bmatrix} + \begin{bmatrix} \overline{b}_1 \\ \overline{b}_2 \end{bmatrix} u(k) + \begin{bmatrix} \overline{w}_1(k) \\ \overline{w}_2(k) \end{bmatrix} \tag{8.4}$$

Then the STO is designed as:

$$\begin{cases} \hat{\overline{x}}_1(k+1) = \overline{a}_{11}\hat{\overline{x}}_1(k) + \overline{a}_{12}\hat{\overline{x}}_2(k) + \overline{b}_1 u(k) + \overline{z}_1 \\ \hat{\overline{x}}_2(k+1) = \overline{a}_{21}\hat{\overline{x}}_1(k) + \overline{a}_{22}\hat{\overline{x}}_2(k) + \overline{b}_2 u(k) + \overline{z}_2 \end{cases} \tag{8.5}$$

$$\hat{\overline{y}}_i(k) = C_i \hat{\overline{x}}(k) \tag{8.6}$$

where $\overline{z}_1 = k_1|\overline{e}_1(k)|^{\frac{1}{2}} sign(\overline{e}_1(k)) + k_2\overline{e}_1(k)$ and $\overline{z}_2 = k_3 sign(\overline{e}_1(k)) + k_4\overline{e}_1(k)$ are the correction terms. C_i is the constant matrix with an appropriate number of channels i. The differences between $C_1, C_2, \ldots, C_n$ are set to simulate the different errors and measurement noises between the channels.

Furthermore, according to Eqs. (8.2)–(8.3) and (8.7)–(8.8), the error of the system states can be written as

$$\overline{e}_1(k+1) = \overline{a}_{11}\overline{e}_1(k) + \overline{a}_{12}\overline{e}_2(k) + \overline{z}_1 - \overline{w}_1(k) \tag{8.7}$$

$$\overline{e}_2(k+1) = \overline{a}_{21}\overline{e}_1(k) + \overline{a}_{22}\overline{e}_2(k) + \overline{z}_2 - \overline{w}_2(k) \tag{8.8}$$

where $\overline{w}(k) = [\overline{w}_1(k), \overline{w}_2(k)]^T$ is bounded, k_1, k_2, k_3, and k_4 are gains that need to be adjusted. Eventually, both $\overline{e}_1(k+1)$ and $\overline{e}_2(k+1)$ will trend to zero, at which time the estimated states are the actual ones. That is, $\hat{\overline{x}}(k) = \overline{x}(k)$.

The error Eqs. (8.7) and (8.8) can be formulated as follows:

$$\overline{e}(k+1) = \overline{P}\overline{e}(k) + \overline{Q}(k) sign(\overline{e}_1(k)) \tag{8.9}$$

where $\overline{P} = \begin{bmatrix} \overline{a}_{11} + k_2\ \overline{a}_{12} \\ \overline{a}_{21} + k_4\ \overline{a}_{22} \end{bmatrix}$, $\overline{Q}(k) = \begin{bmatrix} k_1|\overline{e}_1(k)|^{\frac{1}{2}} \\ k_3 \end{bmatrix}$.

8.3 Network Analysis and Controller Design

In this section, the network-introduced problems, such as communication protocols and network attacks, are modeled, and a controller is designed to stabilize the system under uncertain network conditions.

8.3.1 Communication Protocol

In a NCS, control systems are often connected by a wireless network, while in a wireless network, transmission channels between different nodes are shared. Therefore, data collisions can occur during channel contention.

A communication protocol is required to prevent these problems. A TOD is an effective communication protocol to handle this kind of problem.

As shown in Fig. 1, the measured output signal can be represented as:

$$\overline{y}_k = \left[\left(\overline{y}_k^1\right), \left(\overline{y}_k^2\right), \left(\overline{y}_k^3\right) \dots \left(\overline{y}_k^n\right)\right]^T \in \mathbb{R}^v \tag{8.10}$$

where $\overline{y}_k \triangleq \overline{y}(k)$, and $\overline{y}_k^i$ represents the output signal of the i channel at a discrete time k.

The TOD protocol is proposed to prevent data collision during transmission. The protocol is applied to the feedback channel. After the sensor collects the data, it saves them to the corresponding buffer.

In addition, the buffer also saves the signal returned from the zero-order holds (ZOHs) and the accumulated difference between the current value and the previous one. A selection strategy is used to select the sensor the performs data transmission in the current state.

Based on the TOD protocol described in [5], the TOD protocol used in this paper can be briefly described as follows:

At the beginning of the protocol operation, each sensor channel has the same priority and the same weighting factor. At this time, the selection strategy is based on the difference between the current measured value and the output value of the previous time stored in the buffer, and the channel with the largest difference has the highest priority.

If a channel with the same difference exists, it is selected according to the cumulative value of the difference stored in the buffer, and the highest value has the highest priority. The weighting factor is dynamically adjusted according to the cumulative value.

We set the error of each data buffer as $err_k^i = \overline{y}_k^i - \overline{y}_{k-1}^i$ so that the total buffer can be written as $\overline{E}_k = \left\{err_k^1, err_k^2, err_k^3, \dots, err_k^n\right\}$. The selected node at time k is determined by the variable $\overline{\gamma}_k$,where $\overline{\gamma}_k \in [1, n]$.

We introduce a known positive definite matrix $\overline{W} = diag\{\overline{w}_1, \overline{w}_2, \overline{w}_3, \ldots, \overline{w}_n\}$, where each component $\overline{w}_i$ represents the weight of the sensor i.

According to these, $\overline{\gamma}_k$ can be formulated as:

$$\overline{\gamma}_k = \arg \max_{1 \le i \le n} \left(err_k^i\right)^T \overline{W\Theta}_i err_k^i \tag{8.11}$$

where $\overline{\Theta}_i = \text{diag}\left\{\tilde{\overline{\delta}}(i-1), \tilde{\overline{\delta}}(i-2), \tilde{\overline{\delta}}(i-3), \ldots, \tilde{\overline{\delta}}(i-n)\right\}$, and $\tilde{\overline{\delta}}(\bullet)$ is a Kronecker delta with a value of 0 or 1.

The data transmitted via the TOD protocol are defined as $\tilde{\overline{y}}_k \triangleq \left[\overline{y}_k^1, \overline{y}_k^2, \overline{y}_k^3, \ldots, \overline{y}_k^n\right]$; let $\tilde{\overline{y}}_k^i$ be a component of $\overline{y}_k$, which can be written as:

$$\tilde{\overline{y}}_k = \begin{cases} \overline{y}_k^i, i = \overline{\gamma}_k \\ \tilde{\overline{y}}_{k-1}^i, i \neq \overline{\gamma}_k \end{cases} \tag{8.12}$$

Hence, based on Eqs. (8.11) and (8.12), the signal received by the TOD protocol is expressed as:

$$\tilde{\overline{y}}_k = \overline{\Theta}_{\gamma_k}\tilde{\overline{y}}_k + (I - \overline{\Theta}_{\gamma_k})\tilde{\overline{y}}_{k-1} \tag{8.13}$$

8.3.2 *Model of DoS Attack*

In this model, the output signal of the physical layer is delivered to the cyber layer for transmission. The presence of an attacker is considered in the network transmission, and the attacker attempts to use the DoS attack to ensure that the signal is not transmitted to the remote receiver. The system under the attack function can be abstracted as the following model:

$$\overline{y}_k = C\overline{x}_k + \overline{\Psi\ell}_k \tag{8.14}$$

where $\overline{\Psi\ell}_k$ is the attack term, in which $\overline{\Psi}$ is the attack function and $\overline{\ell}_k$ represents the time when the attack was launched. The attacker launches the DoS attack for the output signal $\overline{y}_k$ to offset the right signal in period $[k_s^d, k_e^d]$, where k_s indicates the start time of the attack and k_e represents the end time. The progress can be formulated as:

$$\begin{cases} \overline{\Psi} = C \\ \overline{\ell}_k = -\overline{x}_k, k \in [k_s^d, k_e^d] \\ \overline{\ell}_k = 0, k \notin [k_s^d, k_e^d] \end{cases} \tag{8.15}$$

We substitute Eq. (8.15) into Eq. (8.14), which can be rewritten as:

$$\begin{cases} \overline{y}_k^w = 0, k \in [k_s^d, k_e^d] \\ \overline{y}_k^r = Gx_k, k \notin [k_s^d, k_e^d] \end{cases} \tag{8.16}$$

We introduce $\overline{\eta}_k$, which indicates the receipt or non-receipt of the observation $\overline{y}_k$ by the feedback controller at time k; the value is 0 or 1. Let $\overline{\Im}_k$ be the set of the data. The estimated observer can be formulated as:

$$\hat{\overline{y}}_k = \mathrm{E}\left(\overline{y}_k \middle| \overline{\Im}_k\right) = \begin{cases} \overline{y}_k, \overline{\eta}_k = 1 \\ \overline{y}_{k-1} = C\hat{\overline{x}}_{k-1}, \overline{\eta}_k = 0 \end{cases} \tag{8.17}$$

8.3.3 Design of Sliding Mode Controller

The error of the output is defined as follows:

$$\overline{\varepsilon}(k) = \overline{y}_k^r - \hat{\overline{y}}_k \tag{8.18}$$

where $\hat{\overline{y}}_k$ is the actual output value of the servo system under DoS attack, and $\overline{y}_k^r$ is the reference input. The discrete-time sliding mode function is defined as follows:

$$\overline{s}(k) = \lambda_1 \overline{\varepsilon}(k) + \lambda_2 \Delta \overline{\varepsilon}(k) \tag{8.19}$$

where $\lambda_1 > 0$, $\lambda_2 > 0$ are the gain parameters, and $\Delta\overline{\varepsilon}(k)$ is defined as $\Delta\overline{\varepsilon}(k) \triangleq \overline{\varepsilon}(k+1) - \overline{\varepsilon}(k)$.

Thus, $\overline{s}(k+1)$ can be formulated as follows:

$$\begin{aligned} \overline{s}(k+1) &= \lambda_1 \overline{\varepsilon}(k+1) + \lambda_2 \Delta \overline{\varepsilon}(k+1) \\ &= (\lambda_1 + \lambda_2)\overline{\varepsilon}(k+1) - \lambda_2 \overline{\varepsilon}(k) \\ &= (\lambda_1 + \lambda_2)(\overline{y}_{k+1}^r - \hat{\overline{y}}_{k+1}) - \lambda_2(\overline{y}_k^r - \hat{\overline{y}}_k) \end{aligned} \tag{8.20}$$

where $\hat{\overline{y}}_k = \overline{y}_k \eta_k + \overline{y}_{k-1}(1 - \eta_k)$.

We introduce a sliding mode reaching law as follows [23]:

$$\overline{s}(k+1) = (1 - qT)\overline{s}(k) - \sigma T \mathrm{sgn} \overline{s}(k) \triangleq \hat{\overline{s}}(k+1) \tag{8.21}$$

So that

$$\hat{\bar{s}}(k+1) = (\lambda_1 + \lambda_2)(\bar{y}^r_{k+1} - \hat{\bar{y}}_{k+1}) - \lambda_2(\bar{y}^r_k - \hat{\bar{y}}_k) \tag{8.22}$$

Thus, the following formulation can be proven from inference:

$$\hat{\bar{y}}_{k+1} = \bar{y}^r_{k+1} - (\lambda_1 + \lambda_2)^{-1}\left[\hat{\bar{s}}(k+1) - \lambda_2(\bar{y}^r_k - \hat{\bar{y}}_k)\right] \tag{8.23}$$

While

$$u_k = \overline{H}^{-1}\left[C^{-1}\hat{\bar{y}}_{k+1} - \overline{G}\hat{\bar{x}}_k\right] \tag{8.24}$$

According to the formulation

$$\hat{\bar{y}}_{k+1} = \bar{y}_{k+1}\bar{\eta}_{k+1} + \bar{y}_k(1 - \bar{\eta}_{k+1}) \tag{8.25}$$

which is

$$\hat{\bar{y}}_{k+1} = \begin{cases} \hat{\bar{y}}_{k+1}, \bar{\eta}_{k+1} = 1 \\ \hat{\bar{y}}_k, \bar{\eta}_{k+1} = 0 \end{cases} \tag{8.26}$$

When $\eta_{k+1} = 1$, the control law is formulated:

$$\begin{aligned} u_k = \overline{H}^{-1}\{C^{-1}\bar{y}^r_{k+1} &- (\lambda_1 + \lambda_2)^{-1}[(1 - qT)\bar{s}(k) \\ &- \sigma T \operatorname{sgn}\bar{s}(k) - \lambda_2(\bar{y}^r_k - \hat{\bar{y}}_k)] - \overline{G}\hat{\bar{x}}_k\} \end{aligned} \tag{8.27}$$

When $\bar{\eta}_{k+1} = 0$, the output value $\hat{\bar{y}}_{k+1} = \hat{\bar{y}}_k$, and the control law equals the control input of the last time.

8.3.4 Stability Analysis

For the stability analysis, the following two theorems are presented.

Theorem 1 *For the discrete-time system* (8.1), *when the reaching law is defined as Eq.* (8.27), *the control strategy of the system is stable, and an arbitrary initial state will approach the switching surface.*

Theorem 2 *For the discrete-time system* (8.1), *when* $\bar{y}^r_k = ref$ *is a constant value, if we appropriately select the parameters* $\lambda_1 > 0$, $\lambda_2 > 0$, *the closed-loop system ensures that* $\lim_{k\to\infty}\left(ref - \hat{\bar{y}}_k\right)$.

Proof We define the $error(k+1) = ref - \hat{\bar{y}}_{k+1}$, according to Eq. (8.25), it can be written as:

$$error(k+1) = ref - [\overline{y}_{k+1}\overline{\eta}_{k+1} + \overline{y}_k(1-\overline{\eta}_{k+1})] \tag{8.28}$$

Case i: $\overline{\eta}_{k+1} = 1$

$$\begin{aligned} error(k+1) &= ref - \hat{\overline{y}}_{k+1} \\ &= ref + C\hat{\overline{x}}_{k+1} \\ &= ref + C(\overline{G}\hat{\overline{x}}_k + \overline{H}u_k) \end{aligned} \tag{8.29}$$

We substitute $\tilde{\overline{y}}_{k+1} = \overline{G}\hat{\overline{x}}_k + \overline{H}u_k$ into Eq. (8.27); the term $\tilde{\overline{y}}_{k+1}$ can be written as follows:

$$\begin{aligned} \tilde{\overline{y}}_{k+1} &= C\{\overline{G}\hat{\overline{x}}_k + \{C^{-1}\overline{y}^r_{k+1} - (\lambda_1+\lambda_2)^{-1}[(1-qT)\overline{s}(k) \\ &\quad - \sigma T\text{sgn}s(k) - \lambda_2(y^r_k - \hat{y}_k)] - G\hat{x}_k\}\} \\ &= y^r_{k+1} - (\lambda_1+\lambda_2)^{-1}[(1-qT)s(k) \\ &\quad - \sigma T\text{sgn}s(k) - \lambda_2(y^r_k - \hat{y}_k)] \end{aligned} \tag{8.30}$$

Thus,

$$error(k+1) = (\lambda_1+\lambda_2)^{-1}[(1-qT)\overline{s}(k) - \sigma T\text{sgn}\overline{s}(k) - \lambda_2(\overline{y}^r_k - \hat{\overline{y}}_k)] \tag{8.31}$$

where $\overline{y}^r_k - \hat{\overline{y}}_k$ equals to $ref - \hat{\overline{y}}_k$ according to Theorem 2.

Hence,

$$error(k+1) = \frac{(1-qT)\overline{s}(k) - \sigma T\text{sgn}\overline{s}(k) - \lambda_2 error(k)}{\lambda_1+\lambda_2} \tag{8.32}$$

Because of $\lambda_1 > 0$, $\lambda_2 > 0$, and $\lim_{k\to\infty} \overline{s}(k) = 0$, and $\lim_{k\to\infty} error(k) = 0$.

Case ii: $\overline{\eta}_{k+1} = 0$

$$\begin{aligned} error(k+1) &= ref + \tilde{\overline{y}}_{k+1} \\ &= ref + C[\overline{G}\hat{\overline{x}}_{k-1} + \overline{H}u_{k-1}] \end{aligned} \tag{8.33}$$

Similarly, we substitute $\tilde{\overline{y}}_{k+1} = \overline{G}\hat{\overline{x}}_{k-1} + \overline{H}u_{k-1}$ into Eq. (8.27), and a similar result can be formulated. This finished the proof.

8.4 Simulation Results

Numerical simulations are carried out to demonstrate the effectiveness of the proposed control methodology. We use the servo system proposed in Eq. (8.1); the

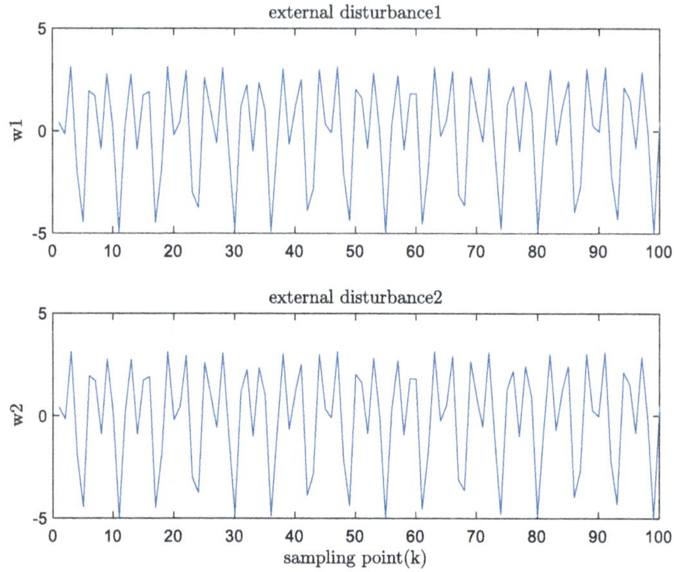

Fig. 8.2 External disturbances injected into the DC motor system

parameters are $\overline{G} = \begin{bmatrix} 0.9647 & -0.0282 \\ 0.0196 & 0.9997 \end{bmatrix}$, $\overline{H} = \begin{bmatrix} 0.0196 \\ 0.0002 \end{bmatrix}$, $C_0 = \begin{bmatrix} -0.2293 \\ 1.4393 \end{bmatrix}^T$, and $C_1 = \begin{bmatrix} -0.2 & 1.4 \end{bmatrix}$. The external disturbance is $\overline{w}_1(k) = 3rand(k) + \sin(k)$, $\overline{w}_2(k) = rand(k) + \cos(2k)$, $rand(k) \in [0, 1]$.

The parameters of the STO are $k_1 = -5, k_2 = -0.6, k_3 = 1, k_4 = 0.2$. The sliding mode controller parameters are $\sigma = 0.001$, $q = 30$, and $T = 0.02$.

The initial conditions are as follows: $\overline{x}(0) = \begin{bmatrix} 200 & 300 \end{bmatrix}^T, \hat{\overline{x}}(0) = \begin{bmatrix} 0 & 0 \end{bmatrix}^T, u(0) = 0, \overline{y}_0^r > 0$. The simulation results are shown as follows.

Figure 8.2 depicts the external disturbances added to the system, which are sinusoidal noises. There are 2 system states: the external disturbances 1 and 2 are injected into $x_1(k)$ and $x_2(k)$ respectively.

Figure 8.3 shows the tracking output at the speed of 1500 r/min during the attack. The system begins to stabilize at about the 60th sampling point (about 1.2 s). The system has multiple output channels but only one channel is selected to transmit data at a time. When the system experiences the maximum error, i.e., the tracking speed deviates from the reference speed, the output channel changes. The dots in Fig. 8.3 represent the time of change. When a DoS attack occurs, the data are lost, but the system quickly regains stability.

Figure 8.4 shows the sliding surface in the simulation at fixed speed. The sliding motion quickly converges to zero and oscillates back and forth around 0. Figure 8.5 shows the control input in the simulation at variable speed. When the system becomes unstable, the controller immediately stabilizes the system.

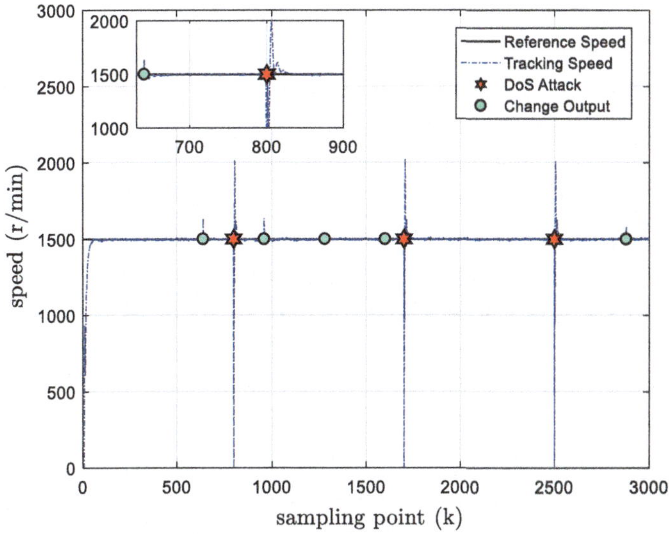

Fig. 8.3 Speed trajectories of simulation at fixed speed

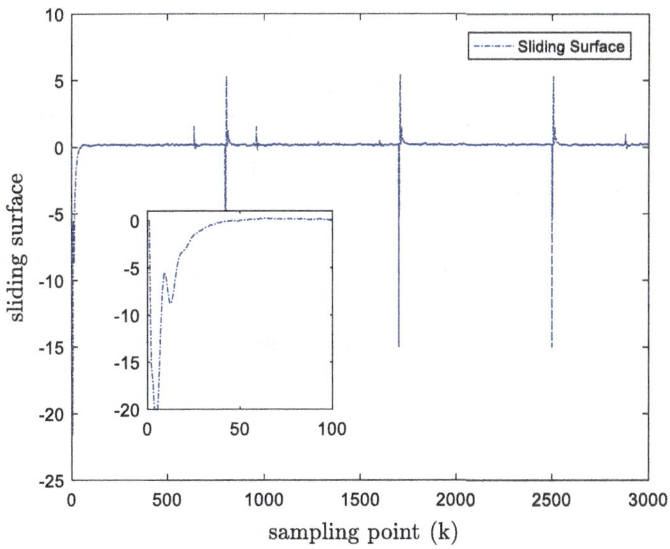

Fig. 8.4 Sliding surface of simulation at fixed speed

A simulation is performed at variable speed. Figure 8.6 shows the tracking output at speeds ranging from 1000 to 2500 r/min during the attack. Similar to Fig. 8.3, at about the 60th sampling point (about 1.2 s), the system starts to stabilize. When the channel changes or a DoS attack occurs, the system quickly converges to a stable

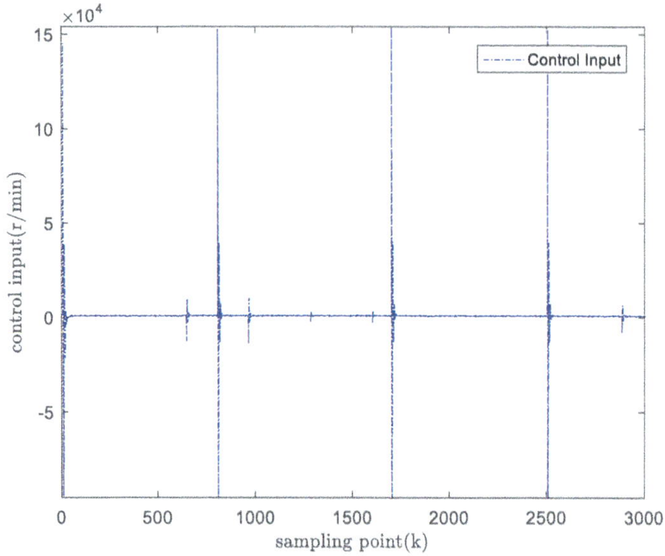

Fig. 8.5 Control input of simulation at fixed speed

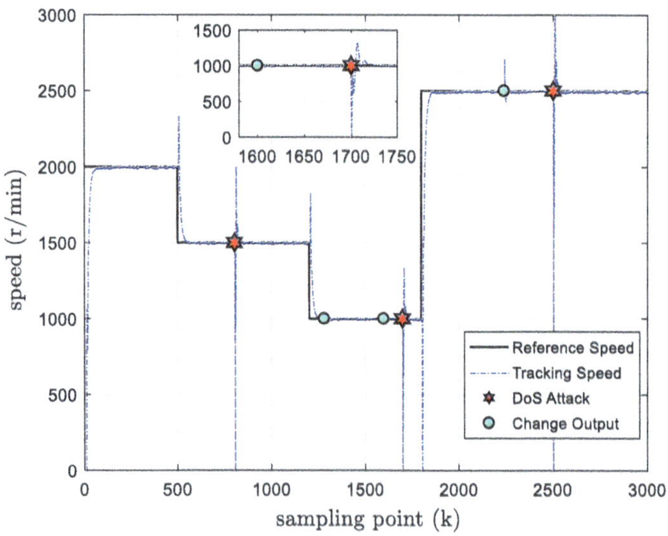

Fig. 8.6 Speed trajectories of simulation at changed speed

state. Figures 8.7 and 8.8 show the sliding surface and control input in the simulation at variable speed. When the channel changes or an attack occurs, the controller consumes energy to prevent the system from becoming unstable.

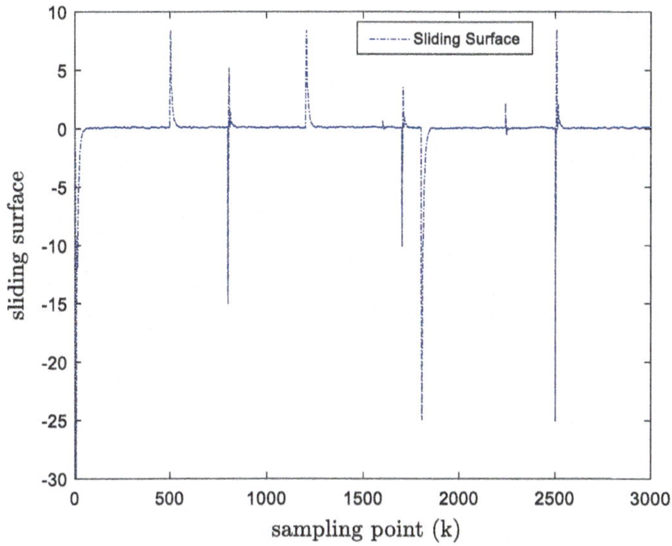

Fig. 8.7 Sliding surface of simulation at changed speed

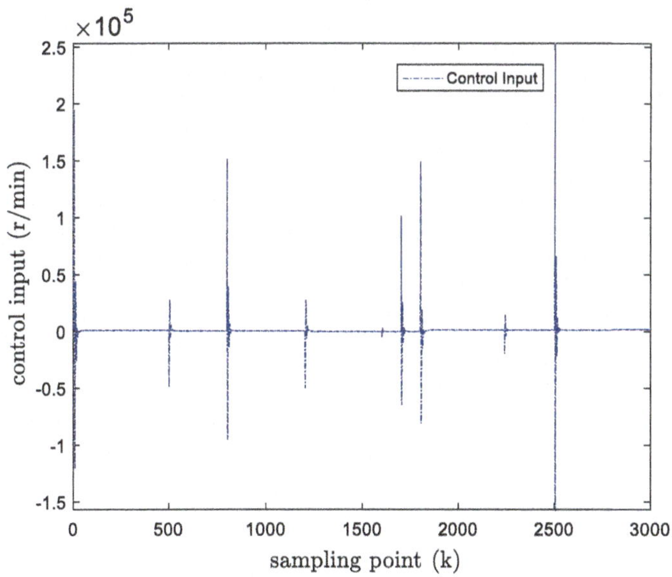

Fig. 8.8 Control input of simulation at changed speed

8.5 Conclusions

This chapter focused on the speed tracking control of a multi-sensor servo system with external interference and DoS attacks. Therefore, data loss and jitter were also introduced. An observer based on the super-twisting algorithm was proposed to observe and estimate external interference. A weight-based TOD protocol was used to select the transmission sensors during DoS attacks. In addition, an SMC method was designed, and the stability of the controller was proved mathematically. Finally, the proposed control method was analyzed using numerical simulations and experiments. The results demonstrated the reliability of the proposed method.

References

1. G. Zhao and C. Hua, Discrete-Time MIMO Reset Controller and Its Application to Networked Control Systems, IEEE Transactions on Systems, Man, and Cybernetics: Systems, vol. 48, no. 12, pp. 2485-2494, Dec. 2018.
2. Y. Li, D. Shi and T. Chen, False Data Injection Attacks on Networked Control Systems: A Stackelberg Game Analysis, IEEE Transactions on Automatic Control, vol. 63, no. 10, pp. 3503-3509, Oct. 2018.
3. Y. Jiang, J. Fan, T. Chai, F. L. Lewis and J. Li, Tracking Control for Linear Discrete-Time Networked Control Systems with Unknown Dynamics and Dropout, IEEE Transactions on Neural Networks and Learning Systems, vol. 29, no. 10, pp. 4607-4620, Oct. 2018.
4. L. Zou, Z. Wang, Q. Han and D. Zhou, Moving Horizon Estimation for Networked Time-Delay Systems Under Round-Robin Protocol, IEEE Transactions on Automatic Control, vol. 64, no. 12, pp. 5191-5198, Dec. 2019.
5. Y. Song, Z. Wang, D. Ding and G. Wei, Robust H2/H∞ Model Predictive Control for Linear Systems with Polytopic Uncertainties Under Weighted MEF-TOD Protocol, IEEE Transactions on Systems, Man, and Cybernetics: Systems, vol. 49, no. 7, pp. 1470-1481, July 2019.
6. Q. Chen, Y. Tan, J. Li and I. Mareels, Decentralized PID Control Design for Magnetic Levitation Systems Using Extremum Seeking, IEEE Access, vol. 6, pp. 3059-3067, 2018.
7. Y. Lu, Adaptive-Fuzzy Control Compensation Design for Direct Adaptive Fuzzy Control, IEEE Transactions on Fuzzy Systems, vol. 26, no. 6, pp. 3222-3231, Dec. 2018.
8. B. Chen, X. Liu and C. Lin, Observer and Adaptive Fuzzy Control Design for Nonlinear Strict-Feedback Systems with Unknown Virtual Control Coefficients, IEEE Transactions on Fuzzy Systems, vol. 26, no. 3, pp. 1732-1743, June 2018.
9. G. Rigatos, P. Siano, A. Melkikh and N. Zervos, A Nonlinear H-Infinity Control Approach to Stabilization of Distributed Synchronous Generators, IEEE Systems Journal, vol. 12, no. 3, pp. 2654-2663, Sept. 2018.
10. A. D. Rosaline and U. Somarajan, Structured H-Infinity Controller for an Uncertain Deregulated Power System, IEEE Transactions on Industry Applications, vol. 55(1), pp. 892–906, Jan.-Feb. 2019.
11. B. Lian, Q. Zhang and J. Li, Integrated Sliding Mode Control and Neural Networks Based Packet Disordering Prediction for Nonlinear Networked Control Systems, IEEE Transactions on Neural Networks and Learning Systems, vol. 30, no. 8, pp. 2324-2335, Aug. 2019.
12. Z. Dong, T. Bao, M. Zheng, X. Yang, L. Song and Y. Mao, Heading Control of Unmanned Marine Vehicles Based on an Improved Robust Adaptive Fuzzy Neural Network Control Algorithm, IEEE Access, vol. 7, pp. 9704-9713, 2019.

13. B. Demirel, E. Ghadimi, D. E. Quevedo and M. Johansson, Optimal Control of Linear Systems with Limited Control Actions: Threshold-Based Event-Triggered Control, IEEE Transactions on Control of Network Systems, vol. 5, no. 3, pp. 1275-1286, Sept. 2018.
14. B. Demirel, A. S. Leong, V. Gupta and D. E. Quevedo, Tradeoffs in Stochastic Event-Triggered Control, IEEE Transactions on Automatic Control, vol. 64, no. 6, pp. 2567-2574, June 2019.
15. S. Zou, A. Mallik, J. Lu and A. Khaligh, Sliding Mode Control Scheme for a CLLC Resonant Converter, IEEE Transactions on Power Electronics, vol. 34, no. 12, pp. 12274-12284, Dec. 2019.
16. Q. Xu, Piezoelectric nanopositioning control using second-order discrete-time terminal sliding-mode strategy, IEEE Transactions on Industrial Electronics, vol. 62, no. 12, pp. 7738-7748, Dec. 2015.
17. A. Ferrara and G. P. Incremona, Design of an integral suboptimal second-order sliding mode controller for the robust motion control of robot manipulators, IEEE Transactions on Control Systems Technology, vol. 23, no. 6, pp. 2316-2325, Nov. 2015.
18. A. Merabet, K. T. Ahmed, H. Ibrahim and R. Beguenane, Implementation of sliding mode control system for generator and grid sides control of wind energy conversion system, IEEE Transactions on Sustainable Energy, vol. 7, no. 3, pp. 1327-1335, July 2016.
19. Q. Xu, Continuous integral terminal third-order sliding mode motion control for piezoelectric nanopositioning System, IEEE/ASME Transactions on Mechatronics, vol. 22, no. 4, pp. 1828-1838, Aug, 2017.
20. J. A. Moreno and M. Osorio, Strict lyapunov functions for the super-twisting algorithm, IEEE Transactions on Automatic Control, vol. 57, no. 4, pp. 1035-1040, Apr, 2012.
21. L. Zou, Z. Wang, H. Gao, Observer-based $H\infty$ control of networked systems with stochastic communication protocol: The finite-horizon case, Automatica, vol. 63, pp. 366-373, Jan. 2016.
22. M. Li and Y. Chen, Robust adaptive sliding mode control for switched networked control systems with disturbance and faults, IEEE Transactions on Industrial Informatics, vol. 15, no. 1, pp. 193-204, Jan. 2019.
23. Meng Li and Yong Chen, Sliding mode predictive tracking control for DC permanent magnet motor in NCSs with random delay and packet dropouts. International Journal of Control Automation & Systems, vol. 15, no. 6, pp. 2723-2732, 2017.

Chapter 9
Terminal Integral Sliding-Mode Tracking Control for Networked Systems with False Data Injection (FDI) Attacks

This chapter investigates the speed tracking control of a network control system (NCS) with external interference and false data injection (FDI) attacks. First, we consider the system model with external interference and FDI attacks. Then, an extended function observer based on a discrete-time sliding function and neural network (NN) is proposed to observe the extended state and suppress the influence of external interference and FDI attacks. On this basis, a novel algorithm combining discrete-time terminal sliding mode control (DTSMC) and super-twist control (STC) is designed to perform closed-loop control of the system. An exponential term and a nonlinear term are included to suppress jitter more effectively. Finally, simulations and experiments prove the validity and feasibility of the proposed theory.

9.1 Introduction

The NCS is a distributed system whose controllers, actuators, sensors, and other components may be located in different areas and are connected through a network [1]. With the rapid development of network and computer technology, the NCS has attracted widespread attention and has been widely used for DC motor control [2], networked vehicles [3], aerospace [4], and other fields. The NCS has many advantages over the traditional bus control system, such as having less wiring, being easily expandable, and not being limited by distance. However, the network results in new problems caused by the uncertainty of network communication. Due to the openness and interconnectivity of the network, attack problems have also arisen. The security of transmitted data requires attention [5].

The network has open access, and all data are transmitted through the network, resulting in a high probability of attacks. Many scholars investigated attacks and security issues. In [6], Niu et al. proposed a novel anomalous intrusion detection method for FID attacks based on a NN. Specific types of attacks causing delays and attacks were predicted. In [7], Mao et al. designed an attack detection algorithm

© The Author(s), under exclusive license to Springer Nature Singapore Pte Ltd. 2022
M. Li et al., *Tracking Control of Networked Systems via Sliding-Mode*,
https://doi.org/10.1007/978-981-16-6514-1_9

that used the characteristic detectability, which effectively solved the problem of zero dynamic attack. Reference [8] analyzed denial-of-service attacks in continuous linear network systems. The authors constructed a novel event-based exchange system and designed a state-based flexible event trigger observer to deal with this type of attack.

Many control solutions for network systems have been proposed to prevent the abovementioned problems. Many advanced control theories exist, and the sliding mode control (SMC) is an important nonlinear control method. It has a simple structure, a fast response speed, and can overcome uncertainty [9]. Many different types of SMC methods have been developed to reduce jitter and improve the control accuracy [10–13]. In [14], a robust adaptive non-singular terminal SMC method was proposed to deal with unknown parameters, uncertain models, and external interference. In [15], a discrete-time fractional-order terminal SMC method for linear motors was proposed to maintain high-precision tracking control. In [16], Xu proposed an SMC strategy based on a terminal sliding mode surface and designed a new second-order discrete-time SMC strategy for precise motion control of piezoelectric-driven nano-positioning devices. By using the feedback of the output value, this strategy was easy to implement and did not require a state observer. In [17], an SMC method based on the discrete-time arrival law effectively controlled disturbances in a discrete-time system. In [18], Liu proposed an adaptive fractional SMC method based on a radial basis function (RBF) NN that was updated online and used in approximately nonlinear systems.

Although many important research results have been obtained, several issues require further research. For example, there are few studies on the use of NNs in discrete systems. Moreover, hybrid control strategies that combine different control methods have not been sufficiently investigated. Motivated by the above studies, we consider the external interference of NCSs and FDI attacks. An extended state sliding mode observer based on an RBF NN is proposed to suppress system state errors caused by external interference and FDI attacks. In addition, a hybrid control strategy using DTSMC and STC is proposed for the effective control of the system. The proposed control method is validated by a numerical simulation, whose results illustrate the effectiveness of the method.

The rest of this article is structured as follows. In Sect. 9.2, the system with external interference is modeled to estimate the interference. In Sect. 9.3, network intrusion is considered, focusing on FDI attacks. A sliding mode observer based on an extended state is proposed to reduce the influence of interference and attacks. An RBF NN is used to reduce the observer's approximation error. In addition, the stability of the method is demonstrated. In Sect. 9.4, the control strategy of the closed-loop system is described, and the new control algorithm consisting of the DTSMC and STC is presented. The convergence of the sliding motion is analyzed. Section 9.5 describes the simulation and experimental results, and Sect. 9.6 summarizes the conclusions of this chapter.

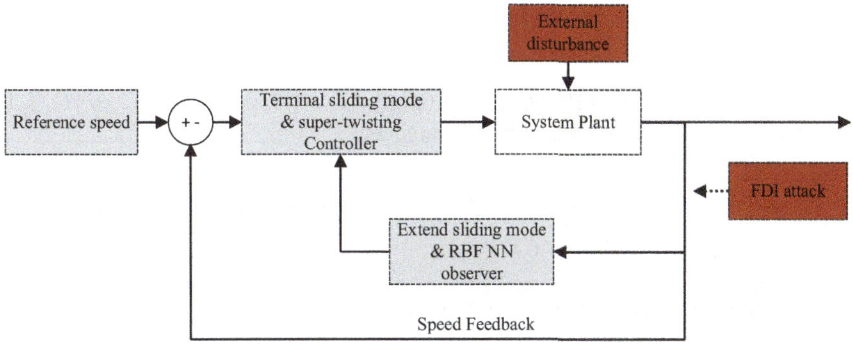

Fig. 9.1 Networked control system with external disturbances and FDI attack

9.2 Problem Formulation

In this section, an NCS with FDI attacks and external interference is modeled. In this model, the actuator is affected by signal interference or other external interference. Moreover, the output signal of the actuator transmitted to the controller is transmitted through the network, and the attacker attempts to use FDI attacks to inject false data into the transmitted data, thereby converting the signal into original data. It there is no appropriate suppression strategy, the system may become unstable. Therefore, we consider a network system with FDI attacks and external interference. Its structure is shown in Fig. 9.1. The model of the system is expressed as follows:

$$\begin{cases} \overline{x}(k+1) = \overline{G}\overline{x}(k) + \overline{H}u(k) + \overline{w}(k) \\ \overline{y}(k) = \overline{C}\overline{x}(k) + \Psi(k) \end{cases} \tag{9.1}$$

where $\overline{x}(k) = [\overline{x}_1(k)\ldots, \overline{x}_n(k)]^T$ and $u(k)$ are the system state and control input, respectively. $\overline{w}(k) = [\overline{w}_1(k)\ldots, \overline{w}_n(k)]^T$ is the disturbance introduced by the network or the parameter drift of the system. $\Psi(k)$ is the attack function, which is used to represent the FDI attack that the attacker launches to offset the output signal $\overline{y}(k)$.

9.3 Design of Augmented State Observer

An augmented state observer based on an RBF NN and sliding mode algorithm is designed to suppress interference of the desired signal. The stability of the observer is demonstrated.

9.3.1 Design of Extended Functional Observer

We set $\tilde{\overline{x}}(k) = [\overline{x}(k), \Psi(k)]^T$, $\tilde{\overline{E}} = [I_n, 0_{n\times p}]$, $\tilde{\overline{G}} = [\overline{G}, 0_{n\times p}]$, and $\tilde{\overline{C}} = [\overline{C}, I_p]$. Then the extended system can be represented as:

$$\begin{cases} \tilde{\overline{E}}\tilde{\overline{x}}(k+1) = \tilde{\overline{G}}\overline{x}(k) + \overline{H}u(k) + \overline{w}(k) \\ \overline{y}(k) = \overline{C}\overline{x}(k) \end{cases} \tag{9.2}$$

The extended functional observer is designed as follows [19]:

$$\begin{cases} \overline{\wp}(k+1) = \overline{M}\overline{\wp}(k) + \overline{N}\overline{y}(k) + \overline{F}u(k) + \overline{v}(k) \\ \hat{\tilde{\overline{x}}}(k) = \overline{\wp}(k) + \overline{J}\overline{y}(k) \end{cases} \tag{9.3}$$

where $\overline{\wp}(k)$ is selected to estimate the extended system states. $\hat{\tilde{\overline{x}}}(k)$ is the estimated value of $\tilde{\overline{x}}(k)$. The matrices $\overline{M}$, $\overline{N}$, $\overline{F}$, and $\overline{J}$ are the parameters of the observer. $\overline{v}(k)$ is used to compensate for the external disturbance.

We define $\overline{R} = \begin{bmatrix} I_n \\ -\overline{C} \end{bmatrix}$ and $\overline{S} = \begin{bmatrix} 0_{n\times p} \\ I_p \end{bmatrix}$ and assume that the pair $\left(\overline{R}\tilde{\overline{G}}, \tilde{\overline{C}}\right)$ is calculable.

Theorem 1 *For the system described in Eq.* (9.2) *and the observer proposed in Eq.* (9.3), *if* $\overline{w}(k) = 0$ *and the pair* $\left(\overline{R}\tilde{\overline{G}}, \tilde{\overline{C}}\right)$ *is calculable, the estimation error* $\tilde{\overline{e}}(k) = \hat{\tilde{\overline{x}}}(k) - \tilde{\overline{x}}(k)$ *will converge to zero.*

The parameters of the observer can be calculated as:

$$\overline{M} = \overline{R}\tilde{\overline{G}} - \overline{Z}\tilde{\overline{C}}, \overline{N} = \overline{Z} + \left(\tilde{\overline{G}} - \overline{Z}\tilde{\overline{C}}\right)\overline{J}, \overline{F} = \overline{RH}, \overline{J} = \overline{S} \tag{9.4}$$

where $\overline{Z}$ is a gain matrix.

Proof We multiply $\overline{R}$ by the first equation in (9.2) and obtain:

$$\overline{R}\tilde{\overline{E}}\tilde{\overline{x}}(k+1) = \overline{R}\tilde{\overline{G}}\tilde{\overline{x}}(k) + \overline{RH}u(k) \tag{9.5}$$

We add $\overline{S}\overline{y}(k+1)$ to Eq. (9.5); according to the second equation of (9.2), Eq. (9.5) can be written as:

$$\overline{S}\tilde{\overline{C}}\tilde{\overline{x}}(k+1) + \overline{R}\tilde{\overline{E}}\tilde{\overline{x}}(k+1) = \overline{R}\tilde{\overline{G}}\tilde{\overline{x}}(k) + \overline{RH}u(k) + \overline{S}\overline{y}(k+1) \tag{9.6}$$

Notice that $\overline{S\tilde{C}} + \overline{R\tilde{E}} = I_{n+q}$. Hence, Eq. (9.6) equals:

$$\tilde{\bar{x}}(k+1) = \overline{R\tilde{G}}\tilde{\bar{x}}(k) + \overline{RH}u(k) + \bar{S}\bar{y}(k+1) \tag{9.7}$$

We define $\bar{\Lambda}(k) = \overline{RH}u(k) + \bar{S}\bar{y}(k+1)$, and system (9.4) can be represented as:

$$\begin{cases} \tilde{\bar{x}}(k+1) = \overline{R\tilde{G}}\tilde{\bar{x}}(k) + \bar{\Lambda}(k) \\ \bar{y}(k) = \tilde{\bar{C}}\tilde{\bar{x}}(k) \end{cases} \tag{9.8}$$

Hence, the observer can be defined as:

$$\hat{\tilde{\bar{x}}}(k+1) = \overline{R\tilde{G}}\hat{\tilde{\bar{x}}}(k) + \bar{\Lambda}(k) + \bar{Z}\left(\bar{y}(k) - \tilde{\bar{C}}\hat{\tilde{\bar{x}}}(k)\right) \tag{9.9}$$

We define $\bar{\wp}(k) = \hat{\tilde{\bar{x}}}(k) - \bar{S}\bar{y}(k)$ so that

$$\begin{aligned} \bar{\wp}(k+1) &= \hat{\tilde{\bar{x}}}(k+1) - \bar{S}\bar{y}(k+1) \\ &= \overline{R\tilde{G}}\hat{\tilde{\bar{x}}}(k) + \bar{\Lambda}(k) + \bar{Z}\left(\bar{y}(k) - \tilde{\bar{C}}\hat{\tilde{\bar{x}}}(k)\right) - \bar{S}\bar{y}(k+1) \\ &= \overline{R\tilde{G}}\hat{\tilde{\bar{x}}}(k) + \overline{RH}u(k) + \bar{S}\bar{y}(k+1) + \bar{Z}\left(\bar{y}(k) - \tilde{\bar{C}}\hat{\tilde{\bar{x}}}(k)\right) - \bar{S}\bar{y}(k+1) \\ &= \overline{R\tilde{G}}\hat{\tilde{\bar{x}}}(k) + \overline{RH}u(k) + \bar{Z}\left(\bar{y}(k) - \tilde{\bar{C}}\hat{\tilde{\bar{x}}}(k)\right) \\ &= \left(\overline{R\tilde{G}} - \overline{Z\tilde{C}}\right)\hat{\tilde{\bar{x}}}(k) + \overline{RH}u(k) + \bar{Z}\bar{y}(k) \\ &= \left(\overline{R\tilde{G}} - \overline{Z\tilde{C}}\right)\left(\bar{\wp}(k) + \bar{S}\bar{y}(k)\right) + \overline{RH}u(k) + \bar{Z}\bar{y}(k) \\ &= \left(\overline{R\tilde{G}} - \overline{Z\tilde{C}}\right)\bar{\wp}(k) + \left(\overline{R\tilde{G}} - \overline{Z\tilde{C}}\right)\bar{S}\bar{y}(k) + \overline{RH}u(k) + \bar{Z}\bar{y}(k) \\ &= \left(\overline{R\tilde{G}} - \overline{Z\tilde{C}}\right)\bar{\wp}(k) + \overline{RH}u(k) + \left[\left(\overline{R\tilde{G}} - \overline{Z\tilde{C}}\right)\bar{S} + \bar{Z}\right]\bar{y}(k) \end{aligned} \tag{9.10}$$

By combining Eq. (9.10) with the observer (9.3), the parameters of the observer can be calculated as Eq. (9.4). This finishes the proof.

We define:

$$\tilde{\bar{e}}(k) = \hat{\tilde{\bar{x}}}(k) - \tilde{\bar{x}}(k) \tag{9.11}$$

According to Eqs. (9.7), (9.9), and (9.11), $\tilde{\bar{e}}(k+1)$ equals

$$\begin{aligned} \tilde{\bar{e}}(k+1) &= \hat{\tilde{\bar{x}}}(k+1) - \tilde{\bar{x}}(k+1) \\ &= \overline{R\tilde{G}}\hat{\tilde{\bar{x}}}(k) + \bar{\Lambda}(k) + \bar{Z}\left(\bar{y}(k) - \tilde{\bar{C}}\hat{\tilde{\bar{x}}}(k)\right) \end{aligned}$$

$$
\begin{aligned}
& - \overline{R\tilde{G}}\tilde{\overline{x}}(k) - \overline{RH}u(k) - \overline{S}\overline{y}(k+1) \\
& = \overline{R\tilde{G}}\hat{\tilde{\overline{x}}}(k) + \overline{RH}u(k) + \overline{S}\overline{y}(k+1) + \overline{Z}\left(\overline{y}(k) - \tilde{\overline{C}}\hat{\tilde{\overline{x}}}(k)\right) \\
& - \overline{R\tilde{G}}\tilde{\overline{x}}(k) - \overline{RH}u(k) - \overline{S}\overline{y}(k+1) \\
& = \overline{R\tilde{G}}\hat{\tilde{\overline{x}}}(k) - \overline{R\tilde{G}}\tilde{\overline{x}}(k) + \overline{Z}\left(\overline{y}(k) - \tilde{\overline{C}}\hat{\tilde{\overline{x}}}(k)\right) \\
& = \left(\overline{R\tilde{G}} - \overline{Z\tilde{C}}\right)\tilde{\overline{e}}(k)
\end{aligned} \tag{9.12}
$$

We define $\Delta\tilde{\overline{e}}(k) = \tilde{\overline{e}}(k) - \tilde{\overline{e}}(k-1)$ and introduce the sliding function as follows:

$$
\overline{s}(k) = q_1\tilde{\overline{e}}(k) + q_2\Delta\tilde{\overline{e}}(k) \tag{9.13}
$$

where $q_1 > 0$, $q_2 > 0$ are the corresponding parameters.

We design the SMC input as follows:

$$
\overline{u}_s(k) = \mu sign(\overline{s}(k)) \cdot |\overline{s}(k)|^{1/2} \tag{9.14}
$$

where $\mu > 0$ is the control parameter, and $sign(f(k))$ represents the sign of function $f(k)$.

The SMC law can be defined as:

$$
\overline{v}(k) = \lambda_1\overline{u}_s(k) + \lambda_2\overline{u}_n(k) \tag{9.15}
$$

where the term $\overline{u}_n(k)$ is an adaptive compensation factor derived from the NN. $\lambda_1 > 0$ and $\lambda_2 > 0$ are the weight parameters, which satisfy $\lambda_1 + \lambda_2 = 1$.

The RBF NN is used to approximate the desired bound of the system uncertainties as:

$$
\hat{\overline{\Phi}}(\hat{\overline{x}}, \hat{\overline{W}}) = \hat{\overline{W}}^T\overline{\varphi}(\hat{\overline{x}}) \tag{9.16}
$$

where $\hat{\overline{x}}$ is the input of the RBF NN, $\hat{\overline{W}} \in R^n$ is the weight vector of the RBF NN, n is the number of nodes in the hidden layer, and the vector $\overline{\varphi}(\hat{\overline{x}}) \in R^n$ is a Gaussian-type function defined element-wise as:

$$
\overline{\varphi}_i(\hat{\overline{x}}) = \exp(-(\hat{\overline{x}} - c_i)^T(\hat{\overline{x}} - c_i)/\sigma_i^2), i = 1, \cdots, n \tag{9.17}
$$

where $c_i \in R^{m\times n}$ and $\sigma_i \in R^n$ denote the center and width of the i th hidden node, respectively; they are predetermined using a local training method [20].

There exists an arbitrary positive constant $\overline{\zeta}_0$ and an optimal constant weight vector $\overline{W}^*$ so that the output of the optimal RBF NN with enough hidden nodes satisfies:

$$\overline{W}^{*}\overline{\varphi}(\hat{\overline{x}}) - \widehat{\overline{\Phi}}(k) = \overline{\zeta}_{f} < \overline{\zeta}_{0} \tag{9.18}$$

where $\overline{\zeta}_f$ denotes the error between the desired upper bound and the estimated upper bound obtained from the NN.

For the design of the NN-based discrete-time sliding mode observer (DSMO), the adjustment of the weight vector, and the analysis of the estimation error convergence, the input of the DSMO can be designed as below.

If the ideal input of $\overline{u}_n(k)$ is $\overline{u}_n^*(k)$, $\overline{u}_n^*(k)$ satisfies the following:

$$\overline{u}_n^*(k) = \overline{W}^{*T} h(z) + \varepsilon(z) \tag{9.19}$$

There exists a positive constant $\overline{W}_m$ and ε_m so that the optimal constant weight vector $\overline{W}^*$ and the approximate error satisfies:

$$\left\| \overline{W}^* \right\| \le \overline{W}_m, \ |\varepsilon(z)| \le \varepsilon_m \tag{9.20}$$

We define the actual NN weight as $\hat{\overline{W}}(k)$ so that the control law designed by the RBF NN is defined as follows:

$$\overline{u}_n(k) = \hat{\overline{W}}^T(k) h(z) \tag{9.21}$$

where $h(z)$ is the output value of the Gaussian-type function, and $z \triangleq \tilde{\overline{x}}(k)$ is the input value of the RBF NN.

According to Eq. (9.21), the following conclusion can be obtained:

$$\overline{u}_n(k) - \overline{u}_n^*(k) = \hat{\overline{W}}^T(k) h(z) - \overline{W}^{*T} h(z) - \varepsilon(z) \tag{9.22}$$

We define $\tilde{\overline{W}} = \overline{W}^* - \hat{\overline{W}}$ so that:

$$\overline{u}_n(k) - \overline{u}_n^*(k) = \tilde{\overline{W}}^T(k) h(z) - \varepsilon(z) \tag{9.23}$$

We define the weight update algorithm as follows

$$\tilde{\overline{W}}(k+1) = \tilde{\overline{W}}(k) - \eta\Big(h(z)\tilde{\overline{e}}(k+1) + \vartheta \hat{\overline{W}}(k)\Big) \tag{9.24}$$

where η and ϑ are positive.

Therefore, Eq. (9.24) can be derived as:

$$\hat{\overline{W}}(k+1) = \hat{\overline{W}}(k) - \eta\Big(h(z)\tilde{\overline{e}}(k+1) + \vartheta\hat{\overline{W}}(k)\Big) \tag{9.25}$$

According to Eqs. (9.15), (9.19), and (9.21), the following applies

$$\tilde{\overline{W}}^T(k)h(z) = \tilde{\overline{e}}(k+1) + \varepsilon(z) \tag{9.26}$$

The estimated states $\hat{\overline{x}}(k)$ and estimated FDI attack $\hat{\psi}(k)$ obtained from the observed can be obtained:

$$\hat{\overline{\overline{x}}}(k+1) = \left(I - \overline{J}\tilde{\overline{C}}\right)^{-1}\left\{\left[\overline{M} + (\overline{N} - \overline{MJ})\tilde{\overline{C}}\right]\hat{\overline{\overline{x}}}(k) + \overline{F}u(k) + \overline{v}(k)\right\} \tag{9.27}$$

We define the matrices as follows to simplify the equation: $\overline{\Theta}_1 = \left(I - \overline{J}\tilde{\overline{C}}\right)^{-1}\left[\overline{M} + (\overline{N} - \overline{MJ})\tilde{\overline{C}}\right], \overline{\Theta}_2 = \left(I - \overline{J}\tilde{\overline{C}}\right)^{-1}\overline{F}, \overline{\Theta}_3 = \left(I - \overline{J}\tilde{\overline{C}}\right)^{-1}$. Since $\hat{\overline{\overline{x}}}(k) = \begin{bmatrix} \hat{\overline{x}}(k) \\ \hat{\psi}(k) \end{bmatrix}$, the extended system can be reshaped as:

$$\begin{gathered}
\hat{\overline{\overline{x}}}(k+1) = \overline{\Theta}_1\hat{\overline{\overline{x}}}(k) + \overline{\Theta}_2 u(k) + \overline{\Theta}_3\overline{v}(k) \\
\begin{bmatrix} \hat{\overline{x}}(k+1) \\ \hat{\psi}(k+1) \end{bmatrix} = \begin{bmatrix} \overline{\Theta}_{11}, \overline{\Theta}_{12} \\ \overline{\Theta}_{13}, \overline{\Theta}_{14} \end{bmatrix}\begin{bmatrix} \hat{\overline{x}}(k) \\ \hat{\psi}(k) \end{bmatrix} + \begin{bmatrix} \overline{\Theta}_{21} \\ \overline{\Theta}_{22} \end{bmatrix}u(k) + \begin{bmatrix} \overline{\Theta}_{31} \\ \overline{\Theta}_{32} \end{bmatrix}\overline{v}(k)
\end{gathered} \tag{9.28}$$

Hence, the original state can be expressed as:

$$\hat{\overline{x}}(k+1) = \overline{\Theta}_{11}\hat{\overline{x}}(k) + \overline{\Theta}_{12}\hat{\psi}(k) + \overline{\Theta}_{21}u(k) + \overline{\Theta}_{31}\overline{v}(k) \tag{9.29}$$

9.3.2 Analysis of Observer Stability

The Lyapunov function is considered as follows:

$$\overline{V}(k) = \tilde{\overline{e}}^2(k) + \frac{1}{\eta}\tilde{\overline{W}}^T(k)\tilde{\overline{W}}(k) \tag{9.30}$$

Thus, $\Delta\overline{V}(k) \triangleq \overline{V}(k+1) - \overline{V}(k)$ equals

$$\begin{aligned}
\Delta\overline{V}(k) &= \tilde{\overline{e}}^2(k+1) - \tilde{\overline{e}}^2(k) + \frac{1}{\eta}\left(\tilde{\overline{W}}^T(k+1)\tilde{\overline{W}}(k+1) - \tilde{\overline{W}}^T(k)\tilde{\overline{W}}(k)\right) \\
&= \tilde{\overline{e}}^2(k+1) - \tilde{\overline{e}}^2(k) + \frac{1}{\eta}\left[\tilde{\overline{W}}(k) - \eta\Big(h(z)\tilde{\overline{e}}(k+1) + \vartheta\hat{\overline{W}}(k)\Big)\right]^T
\end{aligned}$$

$$\times\left[\tilde{\overline{W}}(k)-\eta\left(h(z)\overline{e}(k+1)+\vartheta\hat{\overline{W}}(k)\right)\right]-\frac{1}{\eta}\tilde{\overline{W}}^{T}(k)\tilde{\overline{W}}(k)$$
$$\begin{aligned}&=\tilde{\overline{e}}^{2}(k+1)-\tilde{\overline{e}}^{2}(k)-2\tilde{\overline{W}}^{T}(k)h(z)\tilde{\overline{e}}(k+1)\\&\quad-2\vartheta\tilde{\overline{W}}^{T}(k)\hat{\overline{W}}(k)+\eta h^{T}(z)h(z)\tilde{\overline{e}}^{2}(k+1)\\&\quad+2\eta\vartheta\hat{\overline{W}}^{T}(k)h(z)\tilde{\overline{e}}(k+1)+\eta\vartheta^{2}\hat{\overline{W}}^{T}(k)\hat{\overline{W}}(k)\end{aligned}\tag{9.31}$$

where $|h_i(z)|\leq 1$, $\|h(z)\|\leq l^{1/2}\leq l$, $h^T(z)h(z)=\|h(z)\|^2\leq l$, $i=1,2,...,l$.

So that the following conclusion can be obtained

$$\begin{aligned}&2\vartheta\tilde{\overline{W}}^{T}(k)\hat{\overline{W}}(k)\\&=\vartheta\tilde{\overline{W}}^{T}(k)\left(\hat{\overline{W}}(k)-\overline{W}^{*}\right)+\vartheta\left(\hat{\overline{W}}(k)-\overline{W}^{*}\right)^{T}\hat{\overline{W}}(k)\\&=\vartheta\left(\left\|\tilde{\overline{W}}(k)\right\|^{2}+\left\|\hat{\overline{W}}(k)\right\|^{2}+\tilde{\overline{W}}^{T}(k)\overline{W}^{*}-\overline{W}^{*}\hat{\overline{W}}(k)\right)\\&=\vartheta\left(\left\|\tilde{\overline{W}}(k)\right\|^{2}+\left\|\hat{\overline{W}}(k)\right\|^{2}-\left\|\overline{W}^{*}\right\|^{2}\right)\end{aligned}\tag{9.32}$$

$$\eta h^{T}(z)h(z)\tilde{\overline{e}}^{2}(k+1)\leq\eta l\tilde{\overline{e}}^{2}(k+1)\tag{9.33}$$

$$2\eta\vartheta\hat{\overline{W}}^{T}(k)h(z)\tilde{\overline{e}}(k+1)\leq\eta\vartheta l\left[\left\|\hat{\overline{W}}(k)\right\|^{2}+\tilde{\overline{e}}^{2}(k+1)\right]\tag{9.34}$$

$$\eta\vartheta^{2}\hat{\overline{W}}^{T}(k)\hat{\overline{W}}(k)=\eta\vartheta^{2}\left\|\hat{\overline{W}}(k)\right\|^{2}\tag{9.35}$$

Based on Eqs. (9.32-9.35), Eq. (9.31) can be written as:

$$\begin{aligned}\Delta\overline{V}(k)&\leq\tilde{\overline{e}}^{2}(k+1)-\tilde{\overline{e}}^{2}(k)\\&\quad-2\tilde{\overline{W}}^{T}(k)h(z)\tilde{\overline{e}}(k+1)-\vartheta\left(\left\|\tilde{\overline{W}}(k)\right\|^{2}+\left\|\hat{\overline{W}}(k)\right\|^{2}-\left\|\overline{W}^{*}\right\|^{2}\right)\\&\quad+\eta l\tilde{\overline{e}}^{2}(k+1)+\eta\vartheta l\left[\left\|\hat{\overline{W}}(k)\right\|^{2}+\tilde{\overline{e}}^{2}(k+1)\right]+\eta\vartheta^{2}\left\|\hat{\overline{W}}(k)\right\|^{2}\\&=(-1+\eta(1+\vartheta)l)\tilde{\overline{e}}^{2}(k+1)-\tilde{\overline{e}}^{2}(k)-2\varepsilon(z)\tilde{\overline{e}}(k+1)\\&\quad-\vartheta\left\|\tilde{\overline{W}}(k)\right\|^{2}+\vartheta\left\|\overline{W}^{*}\right\|^{2}+\vartheta(-1+\eta l+\eta\vartheta)\left\|\hat{\overline{W}}(k)\right\|^{2}\end{aligned}\tag{9.36}$$

Equation (9.36) is combined with (9.30) to obtain

$$-2\varepsilon(z)\tilde{\bar{e}}(k+1) \leq g\varepsilon_m^2 + \frac{\tilde{\bar{e}}^2(k+1)}{g} \tag{9.37}$$

where g is a positive parameter.

Moreover,

$$\begin{aligned}\Delta\overline{V}(k) \leq & \left(-1+\eta(1+\vartheta)l+\frac{1}{g}\right)\tilde{\bar{e}}^2(k+1)-\tilde{\bar{e}}^2(k)-\vartheta\left\|\tilde{\overline{W}}(k)\right\|^2 \\ & +\vartheta\left\|\overline{W}^*\right\|^2+\vartheta(-1+\eta l+\eta\vartheta)\left\|\hat{\overline{W}}(k)\right\|^2+\vartheta\overline{W}_m^2+g\varepsilon_m^2\end{aligned} \tag{9.38}$$

when the parameters in Eq. (9.38) satisfy:

$$\begin{cases} g \geq 1 \qquad 0 < \eta(1+\vartheta)l \leq 1-\frac{1}{g} \\ 0 < \eta(1+\vartheta)l \leq 1-\frac{1}{g} \\ 0 < \eta(l+\vartheta) \leq 1 \end{cases} \tag{9.39}$$

the conclusion holds. This finishes the proof.

9.4 Design of Terminal Integral Sliding Mode Tracking Control

In this part, a hybrid SMC method based on the discrete-time SCM and super-twisting algorithm is presented to improve the tracking effect and robustness.

The error of the output is defined as follows:

$$err(k) = \overline{y}^r(k) - \overline{y}(k) \tag{9.40}$$

where $\overline{y}(k)$ is the actual output value of the system under FDI attack, and $\overline{y}^r(k)$ is the reference input.

The discrete-time sliding mode function is designed as follows:

$$\overline{\sigma}(k) = \alpha\Delta err(k) + \beta|err(k-1)|^{q/p} \cdot sign(err(k-1)) \tag{9.41}$$

where $\alpha > 0$ and $\beta > 0$ are the gain parameters. p and q are positive odd numbers satisfying $q/p \in (0, 1)$. $\Delta err(k)$ is defined as $\Delta err(k) \triangleq err(k) - err(k-1)$.

We define the reaching law as follows:

$$\overline{\sigma}(k+1) = \overline{\varsigma}_1\overline{\sigma}(k) \cdot sign(\overline{\sigma}(k)) + \overline{\varsigma}_2 sig^{1/2}\left(\Delta\hat{\psi}(k)\right) + \overline{\varsigma}_3 sign(\overline{\sigma}(k)) \tag{9.42}$$

where $0 < \overline{\varsigma}_1 < 1, 0 < \overline{\varsigma}_2 < 1, \overline{\varsigma}_3 < 0$ are control parameters, and $sig^x(f(k)) \triangleq \|f(k)\|^x \cdot \tanh(f(k))$.

Let $\overline{\sigma}(k+1) = \overline{\sigma}(k)$ so that the equivalent control law can be obtained:

$$\begin{aligned}\overline{u}_{eq}^c(k) = \overline{\Theta}_{21}^{-1}\{\overline{C}^{-1}[\overline{y}^r(k+1) - \frac{1}{\alpha}(\overline{\varsigma}_1\overline{\sigma}(k) \cdot sign(\overline{\sigma}(k)) \\ + \overline{\varsigma}_2 sig^{1/2}\left(\Delta\hat{\psi}(k)\right) + \overline{\varsigma}_3 sign(\overline{\sigma}(k)) + \beta err(k-1)^{q/p}) - err(k)] \\ - \left(\overline{\Theta}_{11}\hat{\overline{x}}(k) + \overline{\Theta}_{12}\hat{\psi}(k) + \overline{\Theta}_{31}\overline{v}(k)\right)\}\end{aligned} \tag{9.43}$$

The equivalent control law is expressed as follows:

$$\overline{u}^c(k) = \lambda_3\overline{u}_{eq}^c(k) + \lambda_4\overline{u}_s^c(k) \tag{9.44}$$

where $\overline{u}_s^c(k)$ is the control input in the form of a super-twisting algorithm, as defined in Eqs. (9.45) and (9.46). $\lambda_3 > 0$ and $\lambda_4 0 > 0$ are the weight parameters, which satisfy $\lambda_3 + \lambda_4 = 1$.

$$\overline{u}_s^c(k) = -\Gamma_1|\overline{\sigma}(k)|^{\frac{1}{2}} sign(\overline{\sigma}(k)) + \tau(k) \tag{9.45}$$

$$\tau(k+1) = -\Gamma_2 sign(\overline{\sigma}(k)) \tag{9.46}$$

where $\Gamma_1 > 0$ and $\Gamma_2 > 0$ are the constant gains, which need to be designed.

The analysis of the sliding motion is stated as follows.

Definition 1 If the following inequation is satisfied, the sliding motion accessibility is proved [21].

$$|\overline{\sigma}(k+1)| \leq |\overline{\sigma}(k)| \tag{9.47}$$

Theorem 2 *If the above-mentioned inequation is satisfied, for the discrete-time system in* (9.1), *the sliding mode function in Eq.* (9.41), *and the equivalent control law in Eq.* (9.44), *the trajectories of the system reach the sliding mode bandwidth within finite steps.*

Assumption 1 We assume that the term containing the estimated FDI attack error satisfies the following condition:

$$\left\|sig^{1/2}\left(\Delta\hat{\psi}(k)\right)\right\| < \|\overline{\varsigma}_3\| \tag{9.48}$$

Proof The proof process can be separated into two steps: sliding motion accessibility and convergence in finite steps.

According to Eq. (9.48), the sliding motion accessibility can be formulated as:

$$\begin{aligned}&\overline{\sigma}(k+1)-\overline{\sigma}(k)\\&=\overline{\varsigma}_1\overline{\sigma}(k)\cdot sign(\overline{\sigma}(k))+\overline{\varsigma}_2 sig^{1/2}\left(\Delta\hat{\psi}(k)\right)\\&\quad+\overline{\varsigma}_3 sign(\overline{\sigma}(k))-\overline{\sigma}(k)\end{aligned}\tag{9.49}$$

When $\overline{\sigma}(k)\geq 0$

$$\begin{aligned}&\overline{\sigma}(k+1)-\overline{\sigma}(k)\\&=(\overline{\varsigma}_1-1)\overline{\sigma}(k)+\overline{\varsigma}_2 sig^{1/2}\left(\Delta\hat{\psi}(k)\right)+\overline{\varsigma}_3\end{aligned}\tag{9.50}$$

where, according to assumption 1

$$\overline{\varsigma}_2\left\|sig^{1/2}\left(\Delta\hat{\psi}(k)\right)\right\|+\overline{\varsigma}_3<\overline{\varsigma}_2\|\overline{\varsigma}_3\|+\overline{\varsigma}_3\tag{9.51}$$

Combined with $0<\overline{\varsigma}_1,\overline{\varsigma}_2<1,\overline{\varsigma}_3<0$, and Eq. (9.51), it can be calculated that $\overline{\sigma}(k+1)-\overline{\sigma}(k)\leq 0$. When $\overline{\sigma}(k)\leq 0$, the same conclusion can be derived so that Eq. (9.47) is satisfied. The sliding motion accessibility has been proved using these steps.

When $\overline{\sigma}(k)\geq 0$, we assume that the sliding motion does not cross the sliding surface when the system approaches the stable state.

$$\overline{\sigma}(1)=\overline{\varsigma}_1\overline{\sigma}(0)+sig^{1/2}\left(\Delta\hat{\psi}(0)\right)+\overline{\varsigma}_3\tag{9.52}$$

where $sig^{1/2}\left(\Delta\hat{\psi}(0)\right)$ contains the estimated attack error at the initial time, which can be regarded as zero.

$$\begin{aligned}&\overline{\sigma}(2)\\&=\overline{\varsigma}_1\overline{\sigma}(1)+\overline{\varsigma}_2 sig^{1/2}\left(\Delta\hat{\psi}(1)\right)+\overline{\varsigma}_3\\&=\overline{\varsigma}_1(\overline{\varsigma}_1\overline{\sigma}(0)+\overline{\varsigma}_3)+\overline{\varsigma}_2 sig^{1/2}\left(\Delta\hat{\psi}(1)\right)+\overline{\varsigma}_3\\&=\overline{\varsigma}_1^2\overline{\sigma}(0)+\overline{\varsigma}_1\overline{\varsigma}_3+\overline{\varsigma}_2 sig^{1/2}\left(\Delta\hat{\psi}(1)\right)+\overline{\varsigma}_3\end{aligned}\tag{9.53}$$

$$\begin{aligned}&\overline{\sigma}(i)\\&=\overline{\varsigma}_1\overline{\sigma}(i-1)+\overline{\varsigma}_2 sig^{1/2}\left(\Delta\hat{\psi}(i-1)\right)+\overline{\varsigma}_3\end{aligned}$$

$$= \overline{\varsigma}_1^i \overline{\sigma}(0) + \overline{\varsigma}_1^{i-1} \overline{\varsigma}_3 + \overline{\varsigma}_1^{i-2} \overline{\varsigma}_3 + \ldots + \overline{\varsigma}_1^0 \overline{\varsigma}_3 + \overline{\varsigma}_1^{i-2} \overline{\varsigma}_2 sig^{1/2}\left(\Delta\hat{\psi}(1)\right) + \overline{\varsigma}_1^{i-3} \overline{\varsigma}_2 sig^{1/2}\left(\Delta\hat{\psi}(2)\right) + \ldots + \overline{\varsigma}_1^0 \overline{\varsigma}_2 sig^{1/2}\left(\Delta\hat{\psi}(i-1)\right) \quad (9.54)$$

According to assumption 1, $\overline{\sigma}(i)$ can be zoomed into

$$\begin{aligned} &\overline{\sigma}(i) \\ &= \overline{\varsigma}_1^i \overline{\sigma}(0) + \overline{\varsigma}_1^{i-1} \overline{\varsigma}_3 + \overline{\varsigma}_1^{i-2} \overline{\varsigma}_3 + \ldots + \overline{\varsigma}_1^0 \overline{\varsigma}_3 \\ &+ \overline{\varsigma}_1^{i-2} \overline{\varsigma}_2 sig^{1/2}\left(\Delta\hat{\psi}(1)\right) + \overline{\varsigma}_1^{i-3} \overline{\varsigma}_2 sig^{1/2}\left(\Delta\hat{\psi}(2)\right) \\ &+ \ldots + \overline{\varsigma}_1^0 \overline{\varsigma}_2 sig^{1/2}\left(\Delta\hat{\psi}(i-1)\right) \\ &\le \overline{\varsigma}_1^i \overline{\sigma}(0) \\ &+ \overline{\varsigma}_1^{i-1} \overline{\varsigma}_3 + \overline{\varsigma}_1^{i-2} \overline{\varsigma}_3 + \ldots + \overline{\varsigma}_1^0 \overline{\varsigma}_3 \\ &+ \overline{\varsigma}_1^{i-2} \overline{\varsigma}_2 \overline{\varsigma}_3 + \overline{\varsigma}_1^{i-3} \overline{\varsigma}_2 \overline{\varsigma}_3 + \ldots + \overline{\varsigma}_1^0 \overline{\varsigma}_2 \overline{\varsigma}_3 \end{aligned} \quad (9.55)$$

Let $\overline{\sigma}(i) = 0, 0 < \overline{\varsigma}_1, \overline{\varsigma}_2 < 1, \overline{\varsigma}_3 < 0$; using Eq. (9.55), a solution can be obatined. $i^* = \lfloor f(\overline{\sigma}(0), \overline{\varsigma}_1, \overline{\varsigma}_2, \overline{\varsigma}_3) \rfloor$ is the finite step. When $\overline{\sigma}(k) \le 0$, a similar conclusion can be obtained. The convergence in finite steps has been proved. This finishes the proof.

9.5 Simulation Results

A numerical simulation is conducted using the system proposed in Eq. (9.1); the parameters are $\overline{G} = \begin{bmatrix} 0.9647 \text{ - } 0.0282 \\ 0.0196\ 0.9997 \end{bmatrix}$, $\overline{H} = \begin{bmatrix} 0.0196 \\ 0.0002 \end{bmatrix}$, and $\overline{C} = \begin{bmatrix} -0.2293 \\ 1.4393 \end{bmatrix}^T$.

The external disturbance $\overline{w}(k) = 2\sin 5k + 3\cos 5k$. The FDI attack point is selected randomly, and the attack size is randomly associated with the output signal. The gain of the observer is $\overline{Z} = [0.07\ -0.1042\,0.3736]^T$, and the corresponding parameters are as follows: $q_1 = 0.03$, $q_2 = 0.22\mu = 0.06$. $\lambda_1 = 0.9$, and $\lambda_2 = 0.1$. The parameters of the sliding controller are: $\alpha = 0.03$, $\beta = 5$, $q/p = 3/5$, $\overline{\varsigma}_1 = 0.11$, $\overline{\varsigma}_2 = 0.95$, $\overline{\varsigma}_3 = -9.8$, $\lambda_3 = 0.65$, $\lambda_4 = 0.35$, $\Gamma_1 = 260$, and $\Gamma_2 = 0.5$. All the initial values are set to zero. The simulation results are shown in Figs. 9.2, 9.3, 9.4, 9.5 and 9.6. Figure 9.2 shows the external disturbance injected into the system states; the disturbance is a mixed sinusoidal signal.

Figure 9.3 depicts the output tracking trajectories at the speed of 1500 rad/min with random FDI attacks. The proposed control method is compared with the discrete-time SMC method without the super-twisting algorithm. The system controlled by the proposed method starts to stabilize at about the 60th point (about 3 s), whereas

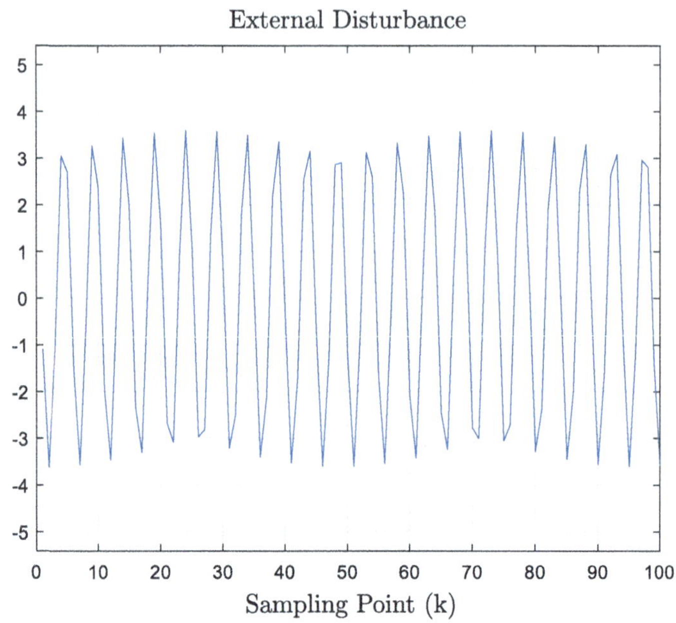

Fig. 9.2 External disturbances injected into the system

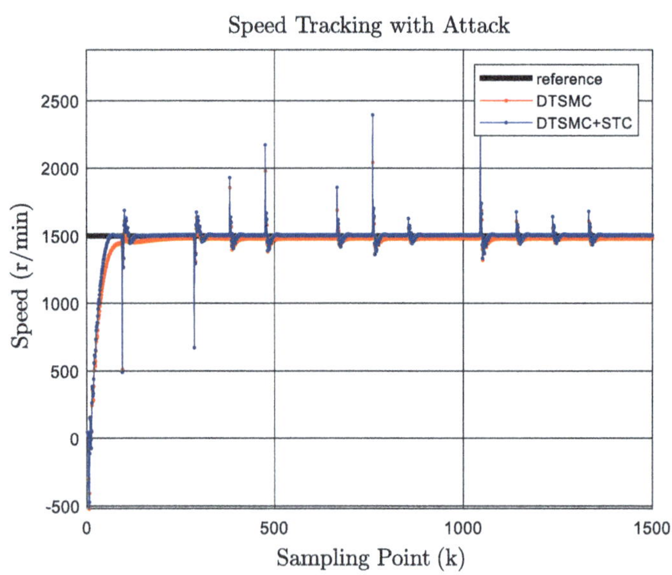

Fig. 9.3 Tracking output of the system at fixed speed

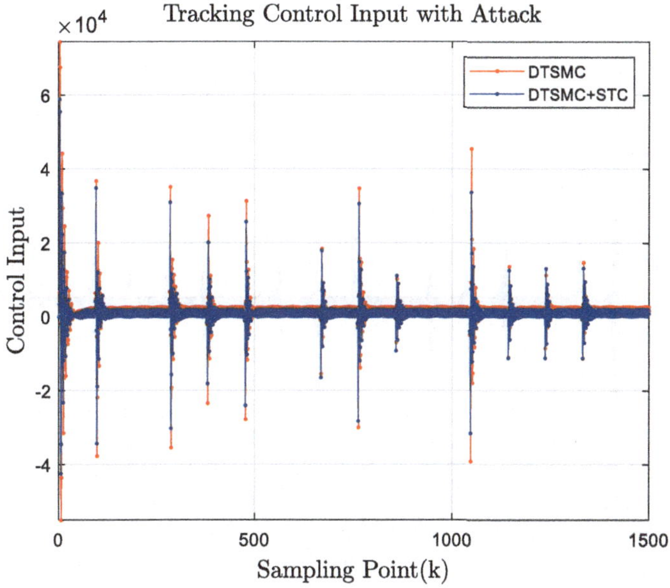

Fig. 9.4 Control input of the system at fixed speed

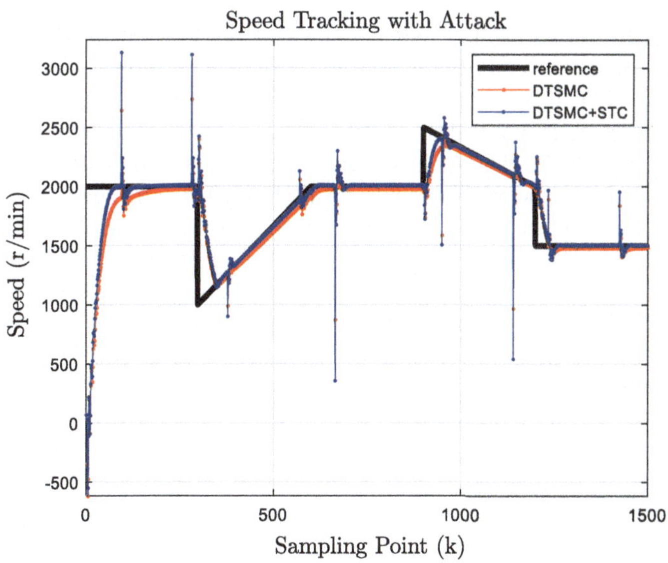

Fig. 9.5 Tracking output of the system at fixed speed

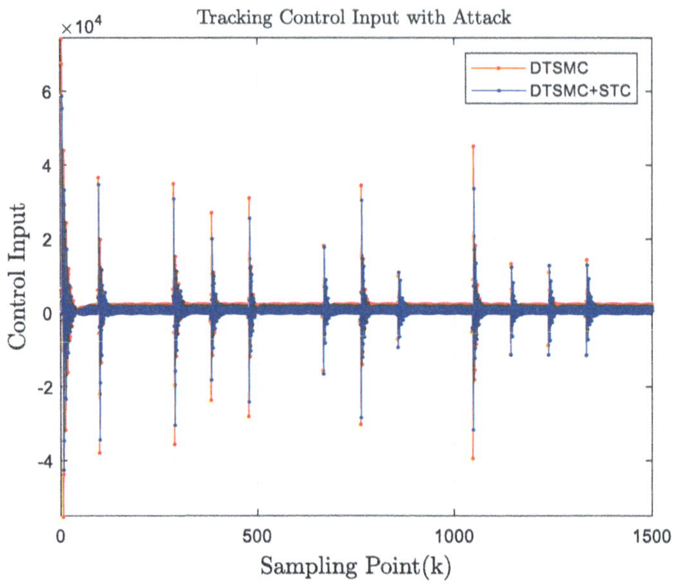

Fig. 9.6 Control input of the system at fixed speed

the discrete-time SMC method becomes stable about 30 points later (about 0.6 s). Meanwhile, the latter method has more static error than the proposed method. Moreover, when the FDI attack occurs, the proposed method becomes stable quicker than the discrete-time SMC method. Figure 9.4 shows the control input of the methods. The proposed method consumes less energy during stabilization after the attack.

The simulation of the tracking control is performed at variable speed.

Figure 9.5 depicts the output tracking trajectories of the system at variable speed for the two control methods. The proposed controller performs significantly better than the discrete-time SMC method. As the speed changes, the proposed control method reacts faster and tracks the attacks more precisely, whereas the discrete-time SMC method has a significant static error. When the FDI attack occurs, the proposed method exhibits slight jitter and stabilizes rapidly.

9.6 Conclusions

In this chapter, an NCS with external disturbance and FDI attacks was investigated, and instability and jitter were introduced. The system was researched and modeled to address these problems. An extended functional observer was established to observe the states and suppress the effect of the FDI attacks and external disturbances. The stability of the observer was demonstrated. Furthermore, a hybrid control method based on discrete-time SMC and a super-twisting algorithm was presented. The reachability of the sliding surface and the convergence in finite steps were proven. Numerical simulations and experiments demonstrated the reliability and effectiveness of the proposed method.

References

1. M. Bahraini, M. Zanon, A. Colombo, and P. Falcone, "Optimal Control Design for Perturbed Constrained Networked Control Systems," IEEE Control Systems Letters, vol. 5, no. 2, pp. 553-558, April 2021https://doi.org/10.1109/LCSYS.2020.3004204
2. M. Ren, J. Zhang, M. Jiang, M. Yu and J. Xu, "Minimum h-infinity Entropy Control for Non-Gaussian Stochastic Networked Control Systems and Its Application to a Networked DC Motor Control System," IEEE Transactions on Control Systems Technology, vol. 23, no. 1, pp. 406-411, Jan. 2015
3. X. Chen, B. Xu, X. Qin, Y. Bian, M. Hu and N. Sun, "Non-Signalized Intersection Network Management with Connected and Automated Vehicles," IEEE Access, vol. 8, pp. 122065-122077, 2020
4. N. Qi, Q. Yuan, Y. Liu, M. Huo and S. Cao, "Consensus Vibration Control for Large Flexible Structures of Spacecraft with Modified Positive Position Feedback Control," IEEE Transactions on Control Systems Technology, vol. 27, no. 4, pp. 1712-1719, July 2019
5. C. Peng and H. Sun, "Switching-Like Event-Triggered Control for Networked Control Systems Under Malicious Denial of Service Attacks," IEEE Transactions on Automatic Control, vol. 65, no. 9, pp. 3943-3949, Sept. 2020
6. H. Niu, C. Bhowmick and S. Jagannathan, "Attack Detection and Approximation in Nonlinear Networked Control Systems Using Neural Networks," IEEE Transactions on Neural Networks and Learning Systems, vol. 31, no. 1, pp. 235-245, Jan. 2020
7. Y. Mao, H. Jafarnejadsani, P. Zhao, E. Akyol and N. Hovakimyan, "Novel Stealthy Attack and Defense Strategies for Networked Control Systems," IEEE Transactions on Automatic Control, vol. 65, no. 9, pp. 3847-3862, Sept. 2020
8. S. Hu, D. Yue, Q. Han, X. Xie, X. Chen and C. Dou, "Observer-Based Event-Triggered Control for Networked Linear Systems Subject to Denial-of-Service Attacks," IEEE Transactions on Cybernetics, vol. 50, no. 5, pp. 1952-1964, May 2020
9. L. Gao, B. Chen and L. Yu, "Fusion-Based FDI Attack Detection in Cyber-Physical Systems," IEEE Transactions on Circuits and Systems II: Express Briefs, vol. 67, no. 8, pp. 1487-1491, Aug. 2020
10. Z. Li, S. Zhou, Y. Xiao and L. Wang, "Sensorless Vector Control of Permanent Magnet Synchronous Linear Motor Based on Self-Adaptive Super-Twisting Sliding Mode Controller," IEEE Access, vol. 7, pp. 44998-45011, 2019
11. J. Li, Y. Niu, "Output-feedback-based sliding mode control for networked control systems subject to packet loss and quantization," Asian Journal of Control, vol. 23, no. 1, pp. 289-297, July 2019
12. P. Gao, G. Zhang, H. Ouyang and L. Mei, "An Adaptive Super Twisting Nonlinear Fractional Order PID Sliding Mode Control of Permanent Magnet Synchronous Motor Speed Regulation System Based on Extended State Observer," IEEE Access, vol. 8, pp. 53498-53510, 2020
13. X. Lin, Y. Wang, Y. Liu, "Neural-network-based robust terminal sliding-mode control of quadrotor," Asian Journal of Control, Dec. 2020.
14. X. Yao, J. H. Park, H. Dong, L. Guo and X. Lin, "Robust Adaptive Nonsingular Terminal Sliding Mode Control for Automatic Train Operation," IEEE Transactions on Systems, Man, and Cybernetics: Systems, vol. 49, no. 12, pp. 2406-2415, Dec. 2019
15. G. Sun, Z. Ma and J. Yu, "Discrete-Time Fractional Order Terminal Sliding Mode Tracking Control for Linear Motor," IEEE Transactions on Industrial Electronics, vol. 65, no. 4, pp. 3386-3394, April 2018

16. Q. Xu, "Piezoelectric nanopositioning control using second-order discrete-time terminal sliding-mode strategy," IEEE Transactions on Industrial Electronics, vol. 62, no. 12, pp. 7738-7748, Dec. 2015
17. A. Bartoszewicz and K. Adamiak, "Discrete-Time Sliding-Mode Control With a Desired Switching Variable Generator," IEEE Transactions on Automatic Control, vol. 65, no. 4, pp. 1807-1814, April 2020
18. N. Liu and J. Fei, "Adaptive Fractional Sliding Mode Control of Active Power Filter Based on Dual RBF Neural Networks," IEEE Access, vol. 5, pp. 27590-27598, 2017
19. M. Li and Y. Chen, "Wide-area robust sliding mode controller for power systems with false data injection attacks," IEEE Transactions on Smart Grid, vol. 11, no. 2, pp. 922-930, Mar. 2020
20. L. Liu, M. Bi, S. Xiao, J. Fang, T. Huang and W. Hu, "OLS-Based RBF Neural Network for Nonlinear and Linear Impairments Compensation in the CO-OFDM System," IEEE Photonics Journal, vol. 10, no. 2, pp. 1-8, April 2018
21. M. Li and Y. Chen, "Robust adaptive sliding mode control for switched networked control systems with disturbance and faults," IEEE Transactions on Industrial Informatics, vol. 15, no. 1, pp. 193-204, Jan. 2019

The manufacturer's authorised representative in the EU is Springer Nature Customer Service Centre GmbH, Europaplatz 3, 69115 Heidelberg, Germany. If you have any concerns regarding our products, please contact ProductSafety@springernature.com

Printed and bound by CPI Group (UK) Ltd, Croydon, CR0 4YY
30/09/2026
02225287-0003